ETOSHA

ETOSHA

DARYL AND SHARNA BALFOUR

STRUIK

Half-title page: The lilacbreasted roller is one of Africa's most startlingly beautiful birds.
Title page: An elephant takes a last drink as the rest of the herd moves away.
Right: The sun sets amid towering rainclouds over a flooded Etosha Pan.
Pages 6/7: A young lion catches the first golden light of dawn.
Pages 8/9: Etosha Pan stretches into the distance in this aerial view over Okerfontein seepage.

*Für John – einen wahren Freund;
und Sean, Nina und Maia
unseren herzlichen Dank.*

Struik Publishers
(a member of The Struik Group (Pty) Ltd)
Struik House
80 McKenzie Street
Cape Town
8001

Reg. No.: 63/00203/07

First published 1992

Text and photographs © Daryl and Sharna Balfour 1992
Map © Loretta Chegwidden 1992

All rights reserved. No part of this publication may be reproduced, stored in a retrieval system, or transmitted, in any form or by any means, electronic, mechanical, photocopying, recording or otherwise, without the prior written permission of the copyright owner(s).

The Publishers are grateful to the following for permission to reproduce copyright material: Virago Press, London and the author's agent, Laurence Pollinger Ltd, London for *West with the Night* by Beryl Markham (page 9), and Panther Books, London for *Going Home* by Doris Lessing (page 26).

Edited by Philippa Parker, Cape Town
Designed by Neville Poulter, Cape Town

Typesetting by Theiner Typesetting Pty (Ltd), Bellville
Reproduction by Chroma Graphics (Overseas) Pte Ltd, Singapore
Printed and bound by Leefung-Asco Printers Ltd, Hong Kong

ISBN 1 86825 135 7

CONTENTS

Acknowledgements   9

Authors' preface   9

Introduction   10

The fulfilment of a dream   14

The wet and the dry   26

The onset of the rains   30

Research and management   38

The big cats   42

Etosha: A wildlife portfolio   49

Photographers' notes   171

Travellers' information   172

Map of Etosha   172-173

References   175

Index   176

ACKNOWLEDGEMENTS

We could never have completed this book without the active and tacit support and assistance of many people. In particular we wish to thank the Government of the Republic of Namibia and the Ministry of Wildlife, Conservation and Tourism for their assistance in granting us permission to live and work in the Etosha National Park. To Allan Cilliers, Raymond du Jardin, Kallie Venzke, Patrick Lane, Dave Gyles, Russell Vinjevold, Duncan Gilchrist, and Wouter Hugo, all in Etosha, we express our sincere appreciation for their advice, co-operation and forbearance.

The generous financial support of our friend, colleague and benefactor, John Matterson of Bushdrifters, and the use of his aircraft for aerial photography, helped us immeasurably, and we are deeply grateful.

Sean, Nina and Maia Beneke of SD General Spares, Mhlume, Swaziland, gave valuable material support in keeping our old 4 x 4 vehicle on the road; Delta Motor Corporation in Port Elizabeth gave generous assistance in obtaining a new Isuzu.

Dave Aronovitz of Hallmark, and Brett Saloner of L. Saul & Co. Ltd in Johannesburg assisted in keeping our battery of Minolta cameras and lenses serviceable, and with the supply of certain equipment from time to time. Creative Colour in Cape Town, under the personal supervision of Dennis Sprong, saw to the processing of much of the film we used.

We wish also to thank our publishers, Struik, in particular Peter Borchert, Eve Gracie, Marje Hemp and Neville Poulter, for their confidence, encouragement, guidance and enthusiasm, and for assigning us to this project in the first place. Our editor, Pippa Parker, and designer, Neville Poulter, are owed much appreciation for the final appearance of this book.

To the many people who helped in so many other ways – carrying film, fetching supplies, passing messages, advising – including Jan and Suzi van der Reep, Nick and Renata Marcodini, Alain Degré, Sylvie Robert, and Chris Charter, we say thank you.

Finally, to our parents, who have always been there for us, we wish to record our heartfelt gratitude for their unstinting love and support over the years; our appreciation is beyond mere words.

AUTHORS' PREFACE

'To see ten thousand animals untamed and not branded with the symbols of human commerce is like scaling an unconquered mountain for the first time, or like finding a forest without roads or footpaths or the blemish of an axe.'
Beryl Markham, *West with the Night*

Once an unspoiled wilderness limited only by natural boundaries, where 'immense numbers of animals' freely roamed, the Etosha of today is barely a quarter of the size of the vast tracts of land that were proclaimed by German Governor F. von Lindequist in 1907. Fences, all-weather roads and luxury air-conditioned accommodation have all contributed towards the taming of this wilderness, and have made it accessible to modern man in his age of indispensable comforts and instant gratification.

But still the most striking aspect of Etosha remains the immense numbers of animals, the huge herds that pepper the plains and crowd the waterholes. These sights will long linger among the most vivid memories of our year in this national park, without doubt one of Africa's finest.

The survival of Africa's wildlife and its habitats are matters of immediate concern to all of us, for as human populations spiral increasingly upwards more and more pressure is exerted upon the wilderness. 'Symbols of human commerce' – the fine roads, luxury restcamps, shops and restaurants – have become an essential part of the management and preservation of Africa's wild creatures and their wild places, for alas, the intrinsic worth of the wild and its beautiful inhabitants is no longer a justification for their protection.

The utilization of a resource as fragile as the ecosystem represented by the Etosha National Park is a weighty responsibility carried by both Government and conservationists, for it is a part of Namibia's heritage, not only for today, but for our children and our children's children. The Ministry of Wildlife, Conservation and Tourism has a proud record in this regard, and we wish them well for the future.

We hope through this portrayal of Etosha and its inhabitants to instil a greater appreciation of such natural resources in all those people who use them, and to encourage a determination to respect life in all its varied and beautiful forms.

DARYL AND SHARNA BALFOUR
HALALI, ETOSHA 1991.

INTRODUCTION

Etosha. The great white place. The place of dry water. The place of mirages. Etosha. The place where you see what is not, and where what is, eludes you.

Whichever interpretation you prefer, in real terms Etosha translates in several other ways: a wildlife preserve without peer; one of the world's last great wildernesses, and a crucial sanctuary for many of Africa's rapidly diminishing bird and game species, including the seriously endangered black rhino, the roan antelope and the rare black-faced impala.

Home to more than 114 mammal and some 340 bird species, Etosha is one of the world's largest national parks, stretching more than 350 kilometres from east to west and covering an area of 22 275 square kilometres. Central to the Park, and from which it derives its name, is the 6 133-square-kilometre pan, once a vast inland sea that gradually dried as climatic changes and topographic movements caused the major river that fed it, the Kunene, to change course and flow into the Atlantic Ocean.

Today Etosha Pan is a stark, skeletal reminder of this former lake, a seemingly endless greenish-white depression of clay, silt and mineral salts which were left behind as the waters evaporated between two and 10 million years ago. Baking and shimmering under the white-gold African sun, the pan has never in living history been totally filled with water, though several tributaries of the Kunene, such as the Ekuma and the Oshigambo rivers in the north-west and the Omuramba Ovambo in the east, still drain into it in years of good rainfall. Such rains cause partial flooding, when vast glistening sheets of water spread over the pan and attract thousand upon thousand of flamingoes, waders and other waterbirds. The birds seem to appear overnight and the pan teems with life until quietly the waters recede, evaporate and once again disappear, leaving the parched, harsh wasteland void of anything but mirages which dance over its scorched, salt-encrusted surface.

Although giving its name to the Park, and unquestionably a central feature of it, Etosha Pan is not 'the Park'. Rather, the surrounding sweetveld savanna plains, founded upon a wide-ranging composite of sands and gravel (the Kalahari Beds) and sustaining vast herds of zebra, wildebeest, elephant, antelope and a myriad other life forms, is what makes for the remarkable wildlife retreat of the Etosha National Park.

THE EARLY EXPLORERS

Although well known and utilized as grazing and hunting lands by resident tribespeople such as the Ovambos, Hereros and nomadic Bushman hunter/gatherers for centuries beforehand, the Etosha Pan and its surrounds were first described for the outside world by Swedish explorer Charles John Andersson and his companion, English scientist Francis Galton (a cousin of Charles Darwin), after they became the first Europeans to travel the area in 1851.

In his book on travel and exploration, *Lake Ngami*, Andersson describes his first encounter with the pan: 'In the course of the first day's journey, we traversed an immense hollow, called Etosha, covered with saline encrustations, and having wooded and well-defined borders. Such places are in Africa designated "salt-pans". The surface consisted of a soft, greenish-yellow, clay soil, strewed with fragments of small sandstone, of a purple tint. In some rainy seasons, the Ovambo informed us, the locality was flooded, and had all the appearance of a lake; but now it was quite dry, and the soil strongly impregnated with salt.'

Andersson and Galton stopped at the site of present-day Namutoni, then known to the Ovambo and Herero pastoralists as *Omutjamatunda*, meaning 'strong waters'. This Andersson described as 'a most copious fountain ... luxuriously overgrown with towering reeds', and recounts that the fountain was swarming with people and cattle, for *Omutjamatunda* was an important cattle post for the Ovambos. The pair and their entourage rested two days before continuing their journey, whiling away the time shooting 'ducks and birds of the grouse kind' which to this day can be found in profusion here.

In the years following Andersson and Galton's visit the spring at *Omutjamatunda* became a popular stop-over for travellers, and in the mid-1870s served as a temporary settlement for a group of Dorsland Trekkers before resuming their journey northwards. These 'thirstland travellers' had left the Transvaal, electing to travel across the harsh interior of the Kalahari Desert in search of a new promised land in Damaraland, because of what they described as 'the un-Godliness of the Pretoria government under President T. F. Burgers, and the freeing of coloured people, as well as other unacceptable laws'. The possibility that the Transvaal might be annexed by the British authorities was another major motivation, although famous Trek leader, Gert Alberts, maintained that they left merely because '*trek was in ons harte*' (trekking was in our hearts).

After months of hardship and deprivation crossing through Bechuanaland (Botswana), Alberts' group reached Rietfontein fountain in the vicinity of what today is Halali on 18 January 1876. Alberts sought permission to remain here temporarily from a succession of native headmen and chieftains, all of whom laid claim to the waterhole. Having been granted the necessary consent, the party remained here in peace and happiness for two years. The tumble-down ruins of stone dwellings seen at Rietfontein today are probably the remains of this settlement, and a memorial dedicated to Johanna Alberts stands in memory to the trek leader's wife – one of several members who died and were buried here.

At the beginning of 1878 the Alberts party left Rietfontein to continue their trek northwards, and a month later arrived at Leeupan where they met up with another group under Jan Greyling. The two joined forces and headed north, encountering the Etosha Pan from its north-eastern side, and describing it as a 'lifeless chalk-flats without water'. They skirted the pan southwards and arrived at the 'strong waters' of Namutoni, where the trek once more came to a halt and reconnaissance parties were sent out to find a suitable place for them to settle. A four-man team travelled westwards on horseback as far as the Atlantic Ocean, but their findings revealed that most of the Kaokoveld was not

Rietfontein is a favourite elephant watering hole.

suitable for human habitation, although they found enough sweet water for the trekkers to stay there temporarily.

After a weary trek westwards through Okaukuejo and present-day Kamanjab, and a 17-month respite at Otjitindua and Kaoko-Otavi, the trekkers finally crossed the Kunene and found a permanent home in Angola.

American trader-adventurer Gerald McKiernan also travelled through what was then German South West Africa in 1876, crossing the Namib Desert and passing through inhospitable country before encountering the fountain at the present-day site of Okaukuejo. His response to the place was one of wonderment: 'All the menageries in the world turned loose would not compare with the sight I saw that day,' he wrote in his diaries, 'it was the Africa I had read of...' McKiernan vividly retold the story of the countryside, writing of 'immense numbers of animals beyond anything I had yet seen. I would scarcely be believed, if I should state that there were thousands of them to be seen at a sight. Gnus in herds like buffalo on the plains, hundreds of zebras, beautiful in their striped coats, springboks by tens of thousands, ostriches, gemsboks and steenboks, hartebeeste and elands...'

Another European to visit the area was Hans Schinz, a Swiss botanist whose interest lay in the flora. Schinz was one of the first to refer to *Omutjamatunda* as 'Amutoni', which is variously recorded in the diaries and writings of other early explorers in the region as 'Great Onamatoni', 'Namatonia' and 'Namatoni'. When in 1897 rinderpest spread from neighbouring Bechuanaland (Botswana), a veterinary control post was established there by the German Government of the day, and the name Namutoni (loosely translated from the Ovambo as 'the high place', for the fountain is situated atop a raised limestone hillock) was settled upon. A similar control post was sited at Okaukuejo, which like Namutoni remained until the epidemic abated. Hereafter, the two posts remained operative as police frontier posts, and in 1901 Okaukuejo was established as a military outpost and a fortified tower was built.

The first Fort Namutoni was constructed in the years 1902-3, about 100 metres from the fountain. Rectangular in form, the fort was constructed from unfired clay bricks; a six-roomed building with battlemented walls and towers at each corner. Soon after its completion it was put to the test. In January 1904 the Hereros rose in revolt against the colonial German Government and the meagre garrison at Fort Namutoni – four men, later joined by three ex-servicemen farming in the district – began to prepare their defences. On 28 January about 500 Ovambos (who joined the Hereros) attacked the fort. The seven defenders held off their attackers until evening, when the Ovambos retreated to re-assess the situation. With their ammunition down to only a handful of bullets, the defenders slipped away under cover of darkness, eventually reaching safety at the town of Grootfontein. The following day, realizing the fort was now undefended, the Ovambo warriors ransacked and razed the building.

Shortly after the rebellion it was decided to rebuild the fort, and construction began in 1905. By the time it was completed the following year, it had gained the reputation of being the most striking fort in German colonial territory. A garrison comprising eight non-commissioned officers and 25 troops was stationed there, their task to prevent smuggling of arms, ammunition and liquor to Ovamboland, as well as to control the movement of migrant labour.

Throughout this time, Etosha and the surrounding areas were regarded as a hunter's paradise, and it was not until 1907 that any interest was shown in conserving the area and its wildlife. On 22 March 1907, the then governor of German South West Africa, Dr F. von Lindequist, out of concern for the rapidly diminishing numbers of wild animals, proclaimed three reserves: Etosha Game Reserve Number 1 was roughly described as an area north of Grootfontein; Etosha Game Reserve Number 2 included all of the present Park as well as most of Kaokoland, and also Damaraland as far north as the Kunene River, west to the coast and as far south as the Hoarusib River; and Etosha Game Reserve Number 3 included large tracts of the Namib Desert to the south.

Von Lindequist's far-sighted actions had the effect of protecting 99 526 square kilometres of countryside, its animals and indigenous plants. There were no fences or physical boundaries, and the game was in no way restricted in its movements. Migration across the game reserve boundaries was not interfered with, and merely meant that the animals were no longer protected, as they were outside of the Park.

Serious interest was shown in researching the animals of the Park when in 1947 the first professional biologist, Mr A. A. Pienaar, was appointed to Etosha. He was succeeded by Dr P. Schoeman in 1951, who showed initiative in proposing the first steps towards game management when he recommended that 1 000 zebras and 500 wildebeest be culled in an attempt to relieve pressure on the grazing lands. In 1954 Mr B. de la Bat was appointed as Etosha's first chief game warden and a nature conservation unit was established, which meant that research and game management would be integral to the running of Etosha. And then in 1974 the Etosha Ecological Research Institute was established with Dr H. Berry as chief biologist.

Other advances and changes in the Park were underway at the same time, notably the development of the tourist function. Okaukuejo was the first indicator of large-scale tourism to come, when in 1952 it was transformed from a police outpost for the control of rinderpest into a restcamp. Namutoni followed in June 1958, the historic fort having been reconstructed true to its original plans which had been found in the Windhoek archives two years previously. The restored splendour of the fort showed no signs of its troubled past and neglect, all of which had taken their toll on this historic landmark declared a national monument in 1950. The third restcamp, Halali, opened to tourists in 1967 following the proclamation of Etosha National Park from its previous title of the Etosha Game Park.

But these years which saw the development of one of Africa's most splendid tourist reserves saw also the redefinition of its boundaries and the reduction of the Park to a mere spectre of the original expanse. This came first in 1928 and then progressively during the years 1958-67, during which time the total areas of Etosha Game Reserves 1 and 3 were deproclaimed. By the early 1970s, following the South African Government's contentious Odendaal Commission, the Park had been reduced in size to its present 22 275 square kilometres, the remainder having been declared 'tribal homelands'.

In 1973, with 75 per cent of Etosha's original area now returned to tribal ownership, the Park's boundaries were formalized by means of a three-metre-high game fence encircling the entire reserve. Although not realized at the time, this ring of steel and wire was to play a major part in the future ecology of the Park, for important migratory routes had been severed and this meant that Etosha would be subjected to ever-increasing grazing pressures. Fenced-in animal populations would now be restricted, forced to remain within man-made boundaries all year round, and the ecology of Etosha would be irrevocably changed.

Ten years later, in March 1983, *National Geographic* magazine reported: 'Even as Etosha shrank, the number of animals within its remaining acreage multiplied tremendously. How? That part was easy. Just add water.'

HEART OF A WILDLIFE PARADISE

While Etosha National Park may not contain Africa's largest concentration of wildlife – a status still held by Tanzania's Serengeti park – it is certainly remarkable for its huge herds, diversity of species and easy accessibility. What makes Etosha extraordinary, however, is the fact that its wildlife is readily visible to even the most casual of visitors, primarily because of a dependence on a limited number of drinking places. As *National Geographic* reported in that same feature (March 1983), 'it may just be possible to enter the lives of a more spectacular array of creatures with greater ease and intimacy here at Etosha than anywhere else on the globe'.

The pan itself holds water for only a few months, and then only in the wettest of years, when torrential rain storms fill small sections of its vast expanse. This water can be as much as two times saltier than sea water, and is thus generally unfit for human and animal consumption. Along the southern shores, however, perennial natural fountains, springs and seepages provide the key to Etosha's burgeoning bounty of wildlife.

Further away from the pan's edge, out across the grassy plains and hidden among acacia and mopane thickets, other waterpoints, both natural and man-made, sustain the teeming herds. Artificial waterpoints are driven by windmill, or by solar or diesel power, and some naturally occurring waterholes are kept topped up in particularly dry years. Such a policy, though, requires careful monitoring and management, for it could cause more problems than it solves. While many of Etosha's animals are to a large extent independent of surface water, extracting what moisture they require from their food, many of the larger mammals must drink on an almost daily basis, thereby limiting their feeding range to within a day's walk of water.

Prior to the fencing of Etosha many of the larger grazing species, such as springbok, wildebeest and zebra, were free to migrate over huge distances as the seasons changed. These animals had distinct dry and wet season ranges which in many instances were beyond the Park's boundaries. Today, however, Etosha's herds have to rely on the food and water available within the Park's fences. Allan Cilliers, Chief Conservation Officer, outlines the problems arising from this situation: 'The fencing and subsequent provision of artificial water supplies can lead to unnatural game concentrations, and, ultimately, to over-grazing ... particularly around these waterpoints. ... To counter this we continually rotate these artificial waterpoints, switching the water supply on or off in a sequence we hope will keep the game moving, encouraging small-scale migrations and thus hopefully preventing over-utilization of any area which may arise were we to leave the water supply constant.'

While many of Etosha's artificial waterholes are conspicuous – windmills or solar panels being clearly visible alongside concrete dishes or troughs – others show few signs of man's hand and appear perfectly natural to the inexperienced eye. These, established as an adjunct to the strictures of fencing, play a vital role in maintaining Etosha's precarious balance of nature, and at the same time enhance game viewing for visitors to the Park.

It is the naturally occurring water, however, upon which the marvel of Etosha is founded, for without the subterranean water supplies and the numerous places where this reaches the surface, little game would have utilized the region, and Reserves 1, 2 and 3 may never have been proclaimed in the early part of the century. Scattered along the edge of the Etosha Pan a series of seepages, or 'contact springs', are formed when the water-laden limestone makes contact with the impervious clay of the pan, forcing the water to the surface. Although generally yielding little water, which in many cases is very brackish, these seeps are of great importance and are often the sites of spectacular congregations of game. Among the most notable of these are Okondeka, Salvadora, Springbokfontein and Okerfontein, all prime game-viewing sites during the dry season months.

Artesian springs also occur in parts of Etosha, usually on ground somewhat higher than the surrounding countryside. These limestone hillocks, such as those at Klein Namutoni and Koinachas in the eastern sector of the Park, were formed over the aeons as water, forced upwards under artesian pressure, evaporated and deposited limestone that was in suspension.

A third type of natural waterpoint found in Etosha is the water-level spring. These are formed in places where the ground surface breaks into the subterranean water level, either when the roof of an underground cave collapses,

A kudu family drinks from the artesian spring at Agab.

or the ground subsides. The two Okevi waterholes are fine examples of the latter, while that at Ngobib developed from an underground cave, the roof of which gave in.

VEGETATION

Other than the parched wasteland of Etosha Pan which dominates the landscape, perhaps the most striking aspect of Etosha National Park is the variation in vegetation cover within its area. Sudden transitions from open grassland to thorn thickets, mopane savanna to tamboti, terminalia and combretum woodland make for a diversity of ecosystems within the Park, and provide food and shelter for a myriad bird and animal species. The pan itself, though described as a saline desert, supports a rich growth of blue-green algae as well as the protein-rich halophytic (salt-loving) grass, *Sporobolus salsus*, which in the dry season offers excellent grazing to huge herds of springbok, gemsbok (oryx) and wildebeest.

Soils immediately adjacent to the pan are brackish, supporting only a few halophytic plants such as the small, shrubby *Suaeda articulata* which covers large tracts in the Salvadora-Charitsaub-Suaeda area near Halali. And on the wider fringes of the Pan the vegetation changes to so-called 'dwarf shrub savanna' which includes the mopane aloe (*Aloe littoralis*) and the widespread water thorn (*Acacia nebrownii*), the species which is responsible for brightening the drab Etosha landscape with brilliant yellow 'puff-ball' flowers from July to September.

Open grasslands rise gradually away from the Pan and sustain a wide variety of perennial and annual grasses, most of which fall into the 'sweetveld' category. These provide protein-rich grazing for huge numbers of herbivores such as springbok, wildebeest and zebra. The winter months see the vast open plains in the Halali region host to herds numbering in the thousands, which are attracted both by the ready availability of water at the numerous contact springs and by the hardy, nutritious grazing here. After the onset of the summer rains, however, most of these herds head west where they utilize the sweet summer grasses in the Grootvlakte area, west of the moringa forest at Sprokieswoud (Enchanted Forest).

The malformed Moringa ovalifolia *trees at Sprokieswoud are the source of many legends.*

Scattered among the grasslands are thorn thickets, or 'thornbush savanna', comprised predominantly of acacia species which provide excellent feed for browsers such as black rhinoceros, giraffe and kudu.

The mopane (*Colophospermum mopane*), characterized by its butterfly-wing foliage, is the dominant tree species and constitutes 80 per cent of the vegetation in Etosha. Both in tree and scrub form, the mopane is a favourite of the Park's nearly 1 750 elephants, which eat the roots, bark, leaves and branches. Interspersed among these trees the purple-pod terminalia (*Terminalia prunioides*), leadwood (*Combretum imberbe*), and red bushwillow (*Combretum apiculatum*) are often found.

The sandveld, or mixed savanna, in the north-eastern sector of the Park owes its existence to deeper soils and a higher annual rainfall than the other areas, and it is here that bigger and more varied tree species may be seen. The graceful and picturesque fan or makalani palm (*Hyphaene ventricosa*) also occurs in this area, seen in impressive form at Fort Namutoni and further east at Two Palms.

Another distinct vegetation type, common in the south-eastern part of the Park, are the tamboti (*Spirostachys africana*) and terminalia forests, which also feature stands of mopane and leadwood trees. These woodlands provide excellent cover for leopards and other smaller predators, and shelter the diminutive Damaraland dik-dik, the smallest antelope occurring in Etosha. They also offer shade and a valuable food source to elephant, giraffe and other browsing animals.

The western sector of Etosha, at present not accessible to visitors, features stark and irregular dolomite formations estimated to be about 50 million years old. Several of these isolated kopjes also occur at Halali, one of which has been included within the camp perimeter. The vegetation of these hills is dominated by the 'ghost tree' or moringa (*Moringa ovalifolia*), the shepherd's tree (*Boscia albitrunca*), common paper-bark (*Commiphora glandulosa*) and scented thorn (*Acacia nilotica*). Two other trees make a visible impact on these dolomite outcrops: the stark large-leaved star chestnut (*Sterculia quinqueloba*) with its white bark, and the slightly less spectacular African star chestnut (*Sterculia africana*).

A similar range of low hills to the west of the Andersson Gate is visible from the water-tower lookout at Okaukuejo. This bears the colourful Herero name of *Ondundozonanandana*, literally translated as 'the place to which the young calves went and from which they never returned', and implying that the range is home to leopards.

Not necessarily a vegetation type, though a distinct part of the Etosha landscape, is the stand of *Moringa ovalifolia* trees which occupies a section of the mopane woodland north-west of Okaukuejo in an area known as 'Sprokieswoud' or the 'Enchanted Forest'. A portion of this stand has been fenced off to protect it from the destructive attentions of animals such as elephants. Moringas usually grow on rocky hillsides and their presence in this flat portion of land remains a mystery, despite botanical investigation into it. Perhaps Dr Hu Berry in his booklet *Etosha National Park* found the answer to this bizarre cluster of trees when he uncovered the legend of the Heikum bushmen, 'age-old inhabitants of the area [who] are less perplexed by this phenomenon, and explain the curious grouping ...: "the Great God in His act of creation had found a place on earth for all animals and plants. His task completed, He realized that He still held a number of moringas in His hand and, not knowing where to plant them, flung them, roots pointing skyward, into their present location."'

THE FULFILMENT OF A DREAM

*E*tosha. The name held an almost mystical appeal for us, for it was a wilderness we had both longed to visit but had never had the opportunity. Together we'd spent almost three years exploring Botswana's wilder reaches, the Okavango, Savuti, Chobe, Nxai Pan, Makgadikgadi, Mabuasehube, and more recently the hills and valleys of Zululand, in the Umfolozi and Hluhluwe game reserves where we had spent time studying rhinos. Kruger was where both of us had spent many childhood holidays. But Etosha was an unfulfilled dream.

The opportunity to encounter and experience Etosha came about when we were approached to write this book. Less than ten weeks later we were struggling to pitch camp in the sparse shade of an acacia thorn, set in a quiet corner of Etosha's main camp, Okaukuejo. It was the beginning of June, and the autumn winds made erecting tents a curse-worthy affair. After nearly three days on the road from Cape Town, the battle to pitch camp in the face of rapidly encroaching darkness was enough to dampen our spirits. Almost. For just then the lions began to roar, a crescendo of heart-stirring, primordial sound. It was as though the Okaukuejo welcoming committee had become aware of our arrival, and were now extending their greetings. For the two of us it meant one thing – we were home.

OKAUKUEJO

Okaukuejo is the biggest camp in Etosha National Park, housing the administrative headquarters of the Park and the Etosha Ecological Research Institute. A camp with superior accommodation facilities, it also has a shop, restaurant, two swimming pools, a post-office and a museum. Its main attraction is a natural spring which draws game from the parched surrounding landscape.

Okaukuejo waterhole has become world-famous. It is conveniently situated on the edge of the restcamp, separated only by a stone wall and overlooked by a paved promenade dotted with trees and benches. Visitors need not even leave the camp to partake of the drama of the bush, and even at night can sit, theatre-style, gazing upon the floodlit arena which hosts all manner of animals, each in turn making its way to the water to quench its thirst.

The lions too are aware of the procession, and as the sun subsides in scarlet splendour, make their own entrance to the scene. Giraffes shuffle nervously in the background, and apparently decide they will drink some other time, tomorrow perhaps. Kudu and springbok turn away, thirsty. And a lone wildebeest canters crazily into the distance. We shiver involuntarily, excitedly, for lions have that effect.

Two male lions, shaggy-maned, sleek and sinuous, crouch to drink while four lionesses and a lone cub drape themselves in golden languor at the base of a gaunt, long-dead leadwood tree.

After a while the roars abate, and peace settles uneasily over the water. Four elephants arrive, disdainful of the dozing lions, and in silence set about satisfying their thirst. The lion cub, about eight months old, stares in fascination, but stays close to the adults. Jackals scurry through the gloom and a pair of dikkops pipe in the night, their musical *pi-pi-pi-peeuw-peeuw-peeuw-pi-pi-pi* in sharp contrast to the fearsome roars of moments before. Then silently from out of the darkness appear two shapes, prehistoric images, relics of a million years past. Perhaps their ancestors knew Etosha in the days when desert was sea? Black rhinos! Gingerly almost, as if fully aware of the threat to their kind, the cow and her half-grown calf make their way to the water, stare short-sightedly about for some seconds, and then drink.

We sit silently, in awe, and watch. Lions, elephants and black rhinoceros together at one waterhole on our first night in Etosha. The dream is real.

JUNE 6. We leave camp early for our first outing in the Park and drive north along the Okondeka track, watching the sun, a molten ball that bulges over the horizon as if loath to start the new day, decant its golden glow over Etosha Pan. We see little other than jackals and springbok, until about a kilometre before Okondeka we encounter a group of lions warming themselves in the morning sun. This, the Okondeka pride, we are to learn in time, are descendants of lions that many years before had eaten several people right beneath the trees now shading this group. 'A nasty bunch', research biologist Lew Scheepers was to tell us some time later. Right now they couldn't appear more indolent, though when I open my door to get a lower camera angle, one of the males immediately tenses into an aggressive crouch, bares his impressive teeth with a threatening snarl, then slinks away. We decide to let the lions be and drive off.

Catching the first warming light of daybreak ...

Later in the morning we stop for brunch beneath a massive colony of sociable weavers, and marvel at the enormous communal nest which appears to house at least 200 birds. The constant activity fascinates us, as birds come and go continually, apparently adding a new sprig of grass or a twig on every return flight. Eventually the nest may become too heavy for the tree, toppling it or, at least, breaking a limb. Other birds roost in these communal nests too: barn owls, barbets and finches, though the pair of pygmy falcons we were told to look out for don't make an appearance.

Driving on we reach Ozonjuitji m'Bari, the western-most point accessible to regular visitors, and site of an artificial waterhole sustained by solar power. In spite of the barren and bleak-looking countryside, great numbers of animals are gathered around the water: several hundred springbok, scattered gemsbok, scores of zebras, wildebeest and a few ostriches, the activities of which provide valuable photographic opportunities, and absorb hours of our time.

There is ample indication of elephant activity in the area too: fresh spoor and droppings about a day old, so we decide to wait until sunset in the hope that they will come. We are not disappointed, for a short while later four bulls make their way across the open grassland towards us. Positioning the vehicle to make best use of the setting sun, we wait. The elephants advance steadily, stopping once with trunks raised to test the wind, all the while their ears flapping gently, until eventually they are drinking less than 20 metres from us. One bull, the biggest and oldest, moves towards us, trunk outstretched, ears spread. He advances steadily and stops about seven metres from us, shakes his head, and flaps his huge ears, a display meant to move us along. We sit silently, and then the elephant relaxes and returns to the water.

On our journey back to Okaukuejo we notice scuffling in the dust alongside the road, and then see the black and silver coat of a honey badger, digging for dinner. We follow as it scurries across the veld, nose to the ground, stopping several times to exhume eggs, grubs and larvae. The badger is still busy when we leave. It has been a good day.

JUNE 7. After spending most of the day at the Okaukuejo waterhole watching some lions make several half-hearted attacks on springbok and other game, we drive out in an easterly direction to Aus waterhole, where we hope to photograph black rhino drinking in the light of a full moon. We arrive at dusk, having stopped en route to watch 14 jackals squabble over the remains of a springbok carcass near Gemsbokvlakte, and immediately go about setting up our cameras and flash equipment. Positioning ourselves in a slight depression about 20 metres from the water, we wait while the darkness falls. In time the moon rises and bathes Aus in silvery light, and we can discern the shape of a rhino walking down a path towards us. Displaying none of the caution we had seen of those at Okaukuejo, this rhino paces straight down to the water, lowers its cumbersome head and begins drinking.

We watch spellbound for several seconds, and then swing into action. At the first flash, the rhino gives a start then stands uncertainly surveying its surroundings. Previous experience has taught us that most wild animals do not react too adversely to camera flashes, and the rhino follows suit, and within moments continues drinking. At the second flash it jumps again, but returns to drinking almost immediately, thereafter showing little reaction at all. After drinking almost continuously for about ten minutes, it walks a short distance away where it stands uncertainly for some time. Just then six hyaenas appear and attract the rhino's attention. With a puff-snort it crashes after them, sending the scavengers scattering in several directions. But its short charge has taken it directly downwind of where we lie watching.

The rhino wheels on the spot, the hyaenas forgotten now that it has caught the scent of its arch enemy, Man. To our surprise it doesn't run away, but begins a curious, mincing advance in our direction. Slowly, one foot at a time, it proceeds. We have no cover and the vehicle is at least 40 metres away, up a rocky slope which is difficult enough walking on, without having to outrun a rhino. We wait. When the rhino is about 10 metres away it stops, cautiously sniffing the breeze. We take the opportunity to photograph it and the flash appears to unnerve it, for it turns and runs about eight paces. Stopping briefly, it begins its advance once more, this time reaching within six or seven metres of us before stopping. Focusing in the bright moonlight, we take another photograph of this rare encounter. The rhino snorts but stands its ground. The animal is now too close – but then something from our time spent studying rhino in Natal recently comes to me and I switch on the camera's motor drive. Raising the camera, I shoot off three frames in quick succession. The rhino spins about on the spot and charges off into the dark, alarmed by the mechanical sounds. He stops about 50 metres away and stares back in our direction, but given this break we make our retreat to the vehicle, and pack up for the night.

JUNE 11. Seven lions lie concealed about Okaukuejo waterhole, alert and motionless, as a lone zebra nonchalantly makes its way down to drink. Suddenly one lioness springs from cover and in a flurry of action the zebra wheels away from the water, spray flying, and begins to run. Two more lionesses join the chase, but the zebra is in full flight and the lions, perhaps not too determined, break off the chase after about 50 metres. Dust hangs in the air, the only reminder of the frenzied moments past. The lions settle once again under a low-hanging tree, shaded and hidden from view.

Several springbok cautiously approach the water, and before reaching it make a swift but half-hearted retreat. A young warthog then appears, jaunty and unwary – until a young cub begins a determined stalk. The disappointment is almost visible on the cub's face when the warthog scampers off. But thirst gets the better of the warthog and soon, accompanied this time by three supporters, it again makes its way towards the water. As the group reaches the

The honey badger scurries across the veld, nose to the ground.

The zebra is in full flight, and the lionesses soon break off the chase.

water's edge a lioness charges, sending stones and dust flying in all directions. The uncomely creatures scatter in disarray and one runs straight into the jaws of a rudely awakened lion, which however welcomes the intrusion, snaps up the offering and proudly walks off with it, the unfortunate warthog squealing and struggling wildly. Even if a bonus, it is not much of a meal and the lion does not share it with the others.

JUNE 17. We rise early and drive out to the waterhole at Olifantsbad where we have been given access to a small hide in the pump-house beneath the windmill. This is used by park rangers for game monitoring purposes. The hide is only a few metres from the water's edge and we have thus an excellent vantage position, abetted by the fact that the windmill and pump are both very noisy, so masking any noises we may make that could possibly alarm the animals. Soon several gemsbok arrive to slake their thirsts, followed by a herd of 26 kudu, an unusually large group. Several of the kudu walk thigh-deep into the water to drink, and it is interesting to see the gemsbok kneeling to reach the water more easily.

Around midday an elephant breeding herd rushes down the slope to the water, several of the 'teenagers' gleefully plunging straight in, rolling and splashing in delight. Among the herd are two of the youngest calves we have seen, hairy and pink behind their ears, still very unco-ordinated and not yet masters of their trunks. At first they try sucking the water into their trunks to emulate the adults, but succeed only in spilling the water before reaching their mouths. Efforts to scoop it up in the curled tip of the trunk also prove unsuccessful. Eventually they give up, and kneel, trunks curled back over their heads, and slurp the water directly into their mouths.

Afterwards, as if irritated with their uselessness, the two youngsters flail their floppy trunks about in the water and are soon immersed, submerging totally on occasion, only the tips of their trunks above the surface. The adults stand drinking from the outlet pipe where water flows into the waterhole, obviously preferring this fresh, flowing supply to the muddy solution in the hole itself.

A second herd of elephants arrives some time later, also with small calves, and we watch a replay of the earlier excitement, this time the entire herd bathing in the waterhole while the other game stands some distance away, watching longingly.

On our return to Okaukuejo shortly after sunset we find two black rhinos at the waterhole, a cow and a sub-adult female calf we have not seen before. In the distance, lions challenge the night, and a jackal cries nearby.

JUNE 22. After spending the morning at the Okaukuejo waterhole watching 'The Old Girl' – a nickname we have given an old lioness that regularly hunts here – we drive north to Okondeka and watch eight ostriches drinking at the seepage out on the pan, and then cut across to m'Bari where we hope to find elephants at sunset. The numbers of animals at the waterhole are far in excess of those we saw on our previous visit: hundreds of springbok, and large groups of gemsbok, wildebeest and zebras. A lone red-billed teal paddles aloofly about the waterhole, rousing the curiosity of several springbok as it passes them by. Three giraffes approach, staring suspiciously at every clump of grass or thorn scrub as they warily circle the hole. Giraffes are at their most vulnerable when drinking, and always do so with utmost circumspection. Eventually they seem satisfied that no lions are lurking nearby, and bend to drink, legs splayed awkwardly apart. Giraffes have special valves in the arteries and veins in their necks which regulate the flow of blood when they bend like this and serve to keep pressure constant during the long haul upwards into a standing position. Despite their abnormally long necks, they have exactly the same number of vertebrae as humans, and indeed as all mammals.

The elephants do arrive, approaching the open grasslands from a mopane belt in the distance. From time to time they stop to dust themselves and feed on the sweet grasses beneath them. Soon they are almost upon us, and then suddenly stop and stare directly towards us. The shrill trumpeting of elephants behind alerts us to a herd more than 30 in number and hastening towards us. They pass within metres of our vehicle and make for the water, where they noisily quench their thirst. We watch as the first three elephants respectfully approach the matriarch of this big group, trunks outstretched in greeting. She responds likewise and, after the greeting ritual is completed, these three join the others at the water, drinking still when we leave.

JUNE 24. It is dark when we leave camp, driving out in the direction of Wolfsnes where we've been told a cheetah with two small cubs had made a kill late

A group of wild dogs huddles together, like domestic pets before a winter fire.

yesterday. Although we know there is little chance of finding her with the kill, we hope that she and her cubs will be somewhere in the vicinity. We scan the countryside, eyes straining in the faint pre-dawn light, then suddenly notice movement in the grass alongside the road ahead. It's a pack of wild dogs, the only seven of this species known to exist in Etosha at present, grouped together in a heap, some on top of others in an effort to derive warmth in the early morning chill. Extinct in most of their former ranges, wild dogs are seriously endangered today due to indiscriminate killing by stock-farmers and hunters in earlier years.

The wild dog's unorthodox methods of hunting have given these animals a poor reputation among those who should know better. They hunt in packs, some of the members harrying the prey closely while the other dogs follow at leisure, taking over the chase when those in front begin to tire. When within reach of their quarry, they bite at its flanks, tearing out pieces of flesh and literally eating it on the run, until eventually the prey falls and is rapidly consumed, sometimes remaining alive for a considerable time.

Early hunters would frequently shoot the wild dogs in disgust and disdain, for their hunting techniques were regarded as cruel and nauseating. The noted early hunter and explorer Frederick Selous, writing in his acclaimed book *A Hunter's Wanderings in Africa*, records watching a single wild dog chase and attack an adult sable antelope, a formidable adversary even for a fully grown male lion. Selous describes the sable as running across the veld at full speed, but 'to no purpose, for the wild dog lying flat to the ground like a greyhound, its bushy tail stretched straight behind it, covered two yards to its one, and came up to it in no time. It gave it just one bite in the flank, and letting go its hold instantly, fell a few yards behind; at the bite the sable antelope swerved toward us, and upon receiving a second, in exactly the same place, turned still more, so that taking the point on which we stood for a centre, both pursuer and pursued had described about a half-circle round us. ... As the wild dog was just going up a third time it got our wind, and instead of again inflicting a bite, stopped dead and looked towards us, whilst about a hundred yards from it the sable antelope also came to a stand. The baffled hound then turned round, and, pursued by Clarkson, made off one way, whilst the sable antelope, delivered from its tormentor, cantered off in another. This is the only time I have ever heard of a wild dog pursuing an animal by itself, especially such a formidable antagonist as a sable antelope bull, which can use its horns with wonderful dexterity. ... Whether in time it would have succeeded in tearing the sable antelope's flank open, and then pulling its entrails out piecemeal, which was its evident intention, I cannot say; but I think it a curious fact, and one well worth noticing, that an African hunting dog is capable of overtaking and attacking single-handed such a powerful animal as a male sable antelope.'

Who would have believed this of the group lying huddled together before us, looking for all the world like domestic pets curled before the hearth of a winter's fire?

Namibian conservation authorities have embarked on an ambitious programme to relocate wild dogs from remote but unprotected areas of the country to Etosha National Park. A recent attempt to stimulate the population of wild dogs in Etosha was carried out with dogs that had been raised in captivity; but this was unsuccessful, the entire pack eventually falling prey to lions or dying of disease.

Leaving the dogs we continue the search for the cheetah and cubs, to no avail, but do encounter a pair of bat-eared foxes prowling the plains on the road towards Adamax. Primarily nocturnal, but occasionally seen in the early mornings and evenings, the foxes have enormous ears which enable them to locate their prey even if it is below the ground. Prey includes insects such as termites, beetles and scorpions, and sometimes also rodents, lizards and birds' eggs and chicks. The fact that farmers frequently shoot these animals to protect their livestock is indeed distressing, as it is physically impossible for a bat-eared fox to kill anything much larger than a rat.

HALALI

Halali, opened to the public in 1968, is the newest of Etosha's restcamps. It is also the smallest, and therefore quietest, of the camps, and is situated in the shadow of 'Twee Koppies', or Helio Hills – so-called because of the siting of a German heliograph unit here in the early days of this century. The area around Halali is one of the most productive for game viewing, as the camp is within easy striking distance of a plethora of excellent waterholes, most of which are natural. We always found the area to be alive with all the major predators and their prey, particularly in the main tourist months through the dry season. The Halali plains are famous for their winter concentrations of zebras, wildebeest, springbok and gemsbok, a sure sign that lions are not far away. We had superb lion encounters as well as numerous cheetah, leopard, elephant and black rhino sightings in the area.

Halali, which derives its name from a German expression for the traditional horn sounded at the end of a hunt, is a delightful camp set among shady trees, with ample grassy lawns. A dolomite outcrop within the camp's perimeter is laid out with several walking trails, and makes for an interesting and relaxing stroll, much needed after a long day in the car. Like the other camps in the Park, Halali has a shop, restaurant, snack kiosk and swimming pool.

JUNE 25. Driving from Okaukuejo to Halali restcamp, which we plan to use as a base for a few days, we pass huge herds of wildebeest and zebra, and nearing Suaeda, a contact spring on the edge of Etosha Pan, we spot four cheetahs. After we've been watching them for a while they move into the

Wildebeest gather on the wide-open plains around Halali.

shade of some thorn scrub. A long wait is rewarded after two hours when the cats move towards us, passing close by the vehicle and then rolling on the road in front of us. Three of the cheetahs are slightly smaller than the other, and we deduce that they are almost fully grown cubs with their mother. They are very playful, and when one flushes a scrub hare from under its feet an excited chase ensues. The hare heads straight towards us, four cheetahs sprinting close behind. The chase then veers off on to the edge of the pan and we are able to catch some of the action on film before the fortunate hare finds safety in a crevice between some rocks. We follow the cheetahs to Suaeda where they drink from the seepage, the four cats looking magnificent against the white salt formations of the spring, but comical too as they struggle to drink without getting their feet wet. We stay with them until darkness falls, but witness no further action.

JUNE 26. We leave Halali before sunrise and drive to Goas fountain, where we find seven lions with bloodied muzzles and bloated bellies. They have obviously just finished feeding and have come to drink before a day of rest. Soon the sun appears, and the lions begin to play. One lioness in particular is in high spirits and teases whichever of the others she comes nearest, biting at tails and pouncing playfully. Another lioness investigates a fallen tree lying at the waterside, and makes an amusing sight as she clambers atop it, battling for balance. After studying the surroundings from her raised vantage point she leaps across the intervening water to join the rest of the pride, the members of which are beginning to move into the mopane trees. Once the lions have disappeared from sight we decide to move on, and follow a track through the bush to another spring further south. Here we find several hundred zebras milling about the water, some of which retreat but are quickly replaced by others. The striking abstract patterns on the hides of these animals are valuable in camouflaging them, particularly in the dark when lions do most of their hunting. The alternately light and dark stripes also help to regulate body temperature. When seen sideways on, the ratio of dark stripes to white background is 3:1. From the rear, however, the ratio of black to white is only 1:3 as there are fewer black stripes here. Using this to their advantage, zebras always stand sideways-on to the sun in the early mornings and in cold weather, the larger and darker body surface here absorbing the sun's heat effectively. At midday the animals stand with their tails to the sun, so exposing a smaller, lighter area of the body to prevent excessive heat absorption.

Later in the morning, at Koinseb fountain, we see our first hartebeest, two groups totalling about 30 individuals which come down to drink. Several herds of the rare black-faced impala also inhabit the area, and come in small groups to water throughout the day.

The sun is past its zenith when we witness an extraordinary performance. A majestic martial eagle, one of the largest of the eagle family, is drinking and bathing in the waterhole nearby, while two blacksmith plovers are aggressively dive-bombing it. The plovers are a fraction of the eagle's size, yet the eagle ducks and weaves at every pass, trying to complete its ablutions in between the attacks of the plucky little plovers, their alarm calls 'tink-tink-tinking' their agitation. The eagle is obviously too cumbersome to evade the attentions of the smaller birds, but its piercing yellow eyes seem to blaze with anger.

Towards sundown we hear the mournful moans of spotted hyaenas, and shortly thereafter several of them make an appearance. They go through elaborate greeting rituals, tails raised, rubbing faces and cowering in the dust. They are clearly all members of the same clan, and eventually we count 11 of them scattered around the spring.

JUNE 27. We have been told of an elephant carcass, a victim of the killer disease anthrax, lying at a remote waterhole. On the way there, travelling down one of the firebreak paths which park rangers have graded throughout Etosha, we encounter an enormous black rhino bull browsing by the roadside. This is our first daylight encounter with a rhino in Etosha, a rare occurrence as the animals are predominantly nocturnal and normally occupy dense thickets where they are difficult to observe. We spend only a short while with him, but shoot a number of photographs before continuing the journey to Gobaub. Here we arrive to find several hundred vultures feasting on the dead elephant, which lies about three metres from the water. Large numbers of jackals are also present, snapping and snarling at the vultures as they too scavenge from the carcass. Many of the vultures are bloated and seemingly incapable of anything more than sitting at the water's edge, eyes glazed.

... she leaps across the water to join the rest of the pride.

The elephant stands at the carcass, picking up the scent of its fallen comrade.

In the middle of the afternoon a herd of 15 elephants appears, with several small calves in the group. It is interesting to see how the mothers with calves nervously skirt the carcass and drink from the opposite side of the waterhole. Shortly, one of the younger cows approaches the carcass, indignantly flapping its ears and waving its trunk to displace the scavenging birds which hop and flap a short distance away. The elephant then stands at the carcass and gently runs its trunk over it, picking up the scent of its fallen comrade. It is a touching moment, and an indication of the close bonds that exist between elephants. In an apparent show of disapproval at the 'desecration' of the remains, the elephant makes several charges at the vultures, chasing them further away. Other members of the herd also come to inspect the carcass now, as if paying their respects to the dead.

After dark four lions come to feed on the dead elephant, and hyaenas lurk and skulk in the background, awaiting their chance. A number of rhinos also appear, and we watch in the moonlight as one of them approaches the carcass. At first the lions stand their ground, snarling determinedly, but back off when it becomes apparent that the rhino is not deterred. However, once the rhino turns its back to leave the carcass, both male lions charge aggressively, and the rhino breaks into a lumbering trot. After about 20 metres, the rhino turns and rushes at the lions, who now turn tail and retreat swiftly. This continues for some time, until the rhino seems to tire of the game and wanders off into the darkness. We watch the lions feed long into the night, and then retire, to the accompaniment of a bushveld symphony of crunching bones, lions snarling, and the occasional whoop of a prowling hyaena.

JUNE 28. The lions are still feeding at sunrise, but as the morning warms up they relinquish their positions to the jackals and vultures. The jackals fill out visibly before our eyes as they gorge themselves. One particular individual who arrived at the scene as a bag of skin and bones feeds until he is so fat that he can barely walk away from the carcass, and only manages the point of the nearest shade where he lies panting in the late morning heat. From time to time a lappetfaced vulture arrives at the feast, and walks imperiously into the fray. This bird clearly commands respect, for all the other scavengers give way. The majority of these are whitebacked vultures, the most common vultures in Etosha and smaller in size, while a few whiteheaded vultures come and go during the day. In the skies above, vultures circle lazily on thermal updrafts, taking a break from the heat on the ground.

Zebras, gemsbok, wildebeest and hartebeest come to drink, but avoid the main pool and the elephant remains, preferring a run-off where the water makes small puddles among the rocks. Towards evening the lions again approach the carcass, scattering all the other scavengers, though by this time most of the vultures have flown to roost in nearby trees. Once again the lions feed all night and only leave at sunrise.

JULY 2. Driving along the road that follows the southern edge of Etosha Pan, we see a cheetah heading towards Suaeda spring. We follow and watch as the cat drinks and then sits on a raised vantage point surveying the scene. Several hundred metres away a group of ostriches on the pan becomes the focus of the cheetah's attention. Immediately, he begins to stalk, his sinuous body close to the ground, his head level with his back. Making use of any available cover, the cheetah makes his way quickly towards the unsuspecting birds, and when about 60 metres from his prey, erupts into action, accelerating within seconds from a crouch to a pace of about 110 kilometres per hour. The ostriches scatter in all directions, perhaps instinctively to confuse the hunter, and pick up pace – they themselves capable of speeds exceeding 70 kilometres per hour – but the cheetah is steadfast and within seconds has one bird in a stranglehold on the dazzling white pan. It takes him several minutes to kill it, reducing its extremely powerful kick to a feeble flutter.

Cheetahs rank low in the hierarchy of predators and are frequently chased off their kills by lion, leopard and hyaena, so it is no surprise that the successful hunter goes about his meal with gusto, frequently looking over his shoulder for scavengers. He eats until his belly bulges, his jaws smeared with drying blood, and leaves only a scattering of feathers in the dirt.

JULY 3. By mid-morning we have had little success, and decide to drive to Agab, a waterhole we are fortunate to have access to, as it is not open to the general public. Driving up the rocky track towards the spring we notice several springbok fixed to the spot, but alert and staring intently in one direction, a sure sign of the presence of predators. We find seven lions – six males and one female – soaking up the morning sun. They are all sub-adults, and are very indifferent to our approach, enabling us to drive very close to the group. After a short while we – and the lions – notice some zebras approaching from the east. The young lioness immediately becomes alert, and after watching for several seconds, begins to stalk them. We watch fascinated as she moves swiftly and silently towards the zebras, her tawny coat blending perfectly with the grass. From time to time she freezes on the spot, and still the zebras have not detected her. But, curiosity getting the better of them, two of the young males decide to get a better vantage position to watch the hunt taking place. They rise and saunter across to a pile of rocks, in full view of the zebras which snort in alarm and run some distance away, now fully alerted to the lions' presence. The disgust with which the lioness regards her siblings is fully evident when she walks disdainfully back past where they lie, and sprawls in the dust some 50 metres from them.

Judging by the age of the lions and the composition of the group, we realize that this is not a pride in itself but part of a larger group, and it comes as no surprise to hear low grunts emanating from the trees a short distance away. All seven lions respond immediately and when an adult lioness walks out into

The seven young lions and the lioness drink their fill.

the open and heads for the water they mob her with obvious glee. With bloodied jowls, evidence of a recent meal, the lioness reaches the water and after drinking her fill disappears into the bush, her gang of seven in tow.

With the lions out of the way other game comes down to drink, and we spend the rest of the morning watching a steady procession of zebras, springbok, gemsbok and hartebeest to and from the water. An unexpected reward is the arrival of a magnificent bateleur eagle which swoops down, its glossy black, brown and white plumage contrasting with its bright red legs, feet and facial skin. The bateleur is one of Africa's most striking eagles, though it is no longer common outside of game reserves. This is partly a result of the actions of farmers who leave poisoned carcasses on their lands in an effort to control predators such as jackal; although an active and capable hunter, the bateleur will readily feed off carrion such as this. After drinking and bathing at the water's edge the bird sits preening in the sun, and then with an effortless beat of mighty wings, takes off and disappears into the afternoon.

Shortly after the bateleur's departure a breeding herd of elephants, totalling more than 40 animals, announces its arrival with a cacophony of rumbling and trumpeting. They charge to the water, disdainful of the animals already there, and chase a herd of zebras in their tracks. The elephants stay at the water for over an hour, drinking and spraying themselves with water but then leave with the same haste that brought them. As the dust settles, a lone hyaena lopes past the waterhole, closely watched by a group of zebras, but does not stop to drink. We return to our campsite as day slips into night, and pack for a trip to the far-western part of the Park tomorrow.

OTJOVASANDU – THE WILD WEST

Etosha is no longer the natural, unspoiled wilderness enthused over by explorers Andersson, McKiernan and others more than 140 years ago. The realities of 20th century economics and the never-ending quest for more land has meant that wildlife preserves need to justify their own existence more positively now in order to avoid further encroachment and possible closure. Fences, roads, water-pumps and restcamps are there because they ensure economic viability, though they destroy the true concept of the wild, and the task of managing, which aims at preserving what is left, often achieves the opposite, destroying the equilibrium that exists in nature.

But there are still places within Etosha that approximate the ideal of a remote and unscarred wilderness, spared the physical signs of development and management. The northern reaches around Natukanaoka Pan and the Ekuma and Oshigambo deltas are such places, closed to visitors and accessible only on horseback or by foot. The far west, at and around Otjovasandu, almost 200 kilometres west of the tourist road at Ozonjuitji m'Bari, is another. Otjovasandu, Herero for 'the place of the young men', is raw and rugged. Burnt-red rocks cut through the soft, silvery savanna, and dolomite kopjes, dark and deterring, echo with the barks of Etosha's only baboon population, and with the whisperings of the spirits of the Heikum – inhabitants of old who hunted in these hills, and left evocative and enduring, primitive art on their stone canvases.

Otjovasandu is a place in which to rejuvenate the soul, where the animals are wild and wary and respond to human intrusion with skittish suspicion or innocent candour. It is inaccessible to tourist visitors as yet, although plans to develop this far-western section of the Park, so physically and spiritually dif-

The country falls away across golden grasslands.

ferent from anything else Etosha has to offer, are underway. Travelling west from Okaukuejo along a dusty and corrugated track known as the 19th parallel (on latitude 19 °S) through featureless mopane woodland, the transition is slow. Rolling hills rise from the landscape so gently at first, almost passing unnoticed until, quite suddenly, the ancient, craggy dolomite ranges lie ahead. The road curves up and through a narrow *poort* and almost abruptly one is in another world.

From the top of the ridge the country falls away across golden grasslands punctuated by dense stands of mopane, stunted boscias and towering combretums, and in the distance purple hills march across the landscape, ending only at the dunes of the Atlantic coast. Here too is where politicians have decreed that the wilderness must end, where fences are erected in disregard of Nature's laws and where the animals are turned back in their tracks. Many of the elephants, oblivious of such proclamations and undeterred by fences, cross the boundaries at will – at their peril, however. There is a ranger outpost at Otjovasandu, near the ancient spring which yields some of the sweetest water in Etosha. Incumbents confront the difficult task, among others, of policing the boundaries, beyond which lie tribal lands belonging to Hereros, Damaras and Ovambos, the 'tribal homelands' created by the South African Government prior to Namibia's independence. It is difficult not to be wistful in Otjovasandu, remembering that when Etosha Game Reserve Number 2 was proclaimed in 1907 its boundary to the west was the sea and to the north the Kunene River; it is here more than anywhere else that one can see what has been lost through time and 'progress'.

Located in this south-western corner at Otjovasandu is the Kaross sanctuary for endangered species, a 15 000-hectare enclave where rarer game species, such as roan antelope, eland, black-faced impala, mountain zebra and black rhino, receive special attention. Established in the 1970s, Kaross is a haven of tranquillity, set amid some of Etosha's most beautiful countryside. Granite outcrops, like a misplaced Stonehenge, jut from the landscape and ethereal white-barked *Sterculia quinqueloba* trees stand like sentinels atop dolomite kopjes. Dry river courses twist through deep gorges and diminutive klipspringers stand watch among the boulders of the towering crags.

Kaross is in fact managed more as a farm for endangered species than as a game reserve, and the landscape here is dotted with artificial waterholes to cater for the animals. It is possible that in time the sanctuary will be the arena

THE FULFILMENT OF A DREAM

We camped in the shadows of granite boulders.

A nocturnal encounter between two black rhinos.

for the Park's first guided walking trails. A better place could not be found for here there is a sensitivity and serenity not readily found in the harsh, featureless expanse in the eastern part of the Park.

During our first visit to Kaross we made camp in the shadow of a jumble of granite boulders and rambled through valleys and gorges, and along dry riverbeds.

JULY 13. Waiting quietly, hidden among the boulders overlooking Zebrapomp waterhole, we watch as several Hartmann's mountain zebras nimbly traverse a rocky hillside. A pair of Monteiro's hornbills glide across the clearing, settle on a fallen tree-trunk and forage for grubs and beetles. Later a group of black-faced impala, delicate, though somewhat heavier than the common impala found elsewhere in southern Africa, feed closer by, several of the rams taking turns at standing sentry when the group comes for water. Towards evening, when the shadows lengthen and soften the landscape, a small flock of rosy-faced lovebirds, jewel-like in their pinks, blues and greens, screech overhead, then settle at the waterside. As darkness washes over Kaross, a leopard, unseen, coughs quietly and baboons chatter in alarm. The staccato calls of barking geckoes echo over the rocks, and a nightjar begins its haunting lovesong.

And somewhere in the darkness, a black rhino chuffs, like an asthmatic steam engine shunting in the night.

During the night several rhinos arrive to drink. We are able to stalk quite close to observe and photograph them, and we witness a frightening encounter between two individuals as they spar with their horns, squealing, grunting and puffing.

JULY 16. Taking cover in a makeshift hide beneath a candlepod acacia at a waterhole called Karosshoek, we watch the game arriving at the nearby water trough. Roan antelope, seriously endangered throughout southern Africa, are feeding nearby and throughout the day small family groups come to drink. We suppress our chuckles when a warthog, attempting to mount a female, topples into the water, hauling himself out with visible embarrassment. Later another warthog comes to wallow in the mud patch alongside the overflow pool – created for just this purpose. Many of Africa's game species wallow in mud to protect themselves from skin parasites such as ticks and biting flies, and as a means of regulating their body temperature.

Some time later a group of nine kudu bulls arrives to drink. The formation of such 'bachelor groups' is typical out of the breeding season, the bulls joining up with the females only during the rut. In the late afternoon a herd of eland makes a cautious approach, shy and skittish. Eventually the eland reach the water, giving us the opportunity to marvel at their huge but stately form. Africa's largest antelope species, an adult eland bull can weigh in excess of 850 kilograms, yet for all its bulk can effortlessly clear a two-metre fence in a single leap. Usually fawn-grey in colour, large old bulls take on a deep, lustrous bluish colour which explains the derivation of the term 'blou bul' used by the early hunters.

And all the while thousands of red-billed queleas drink in continuous droves; the birds enshroud a small scrub acacia alongside the water trough and appear to drink in relays of ordered confusion, hundreds of birds swarming from the branches down to the water and back every few seconds.

Later that evening as we sit at the campfire in the darkness, listening to the myriad profound whisperings of the African bush, we reflect on the words of Swiss psychologist Carl Jung, wondering if he too had experienced a place like Karosshoek when he wrote in 1925 that 'Africa has the stillness of the eternal beginning'.

Black-faced impala ewes drink from the water trough at Zebrapomp.

NAMUTONI: THE PLACE OF THE LEOPARD

Namutoni is the most picturesque of Etosha's restcamps, situated on a slight rise overlooking the surrounding countryside and flanked by towering fan palms. It is also the site of the historic fort, a relic of German colonial days, now fully restored. The startling white battlemented walls and lookout towers impart a romantic aura to the place, but most memorable is the view from atop the flag tower at sunset, when *The Last Post* is sounded to the lowering of the Namibian flag.

Situated in the eastern sector of the Park, Namutoni translates from Ovambo as 'the high place', but to us it meant leopards. The strikingly different vegetation here, ranging from tamboti and terminalia woodland to mixed tree and shrub savanna, seems to provide the ideal habitat for these secretive and elusive predators, and it was around Namutoni that we had our best sightings of these cats.

A great number of attractive and rewarding waterholes dot the area, most within close proximity of the camp and each other, making game viewing here almost as easy as sitting at the Okaukuejo waterhole. The nearby Fischer's Pan, when filled with water after the rainy season, is a birdwatcher's delight, teeming with flamingoes, pelicans and diverse waders and waterfowl. Namutoni is also the area in which you are most likely to see the diminutive Damara dik-dik, Etosha's smallest antelope, particularly on the aptly named Bloubokdraai (Dik-dik Drive) which never fails to provide a sighting of them.

JULY 22. The water in Fischer's Pan is shrinking every day, and now only 300 or so flamingoes and a scattering of pelicans remain. Soon they too will be gone, only to return at the end of the year after the summer rains have filled the pan once more. Nobody really knows where the flamingoes go to when

Namutoni Fort is a striking reminder of former colonial times.

they leave Etosha, or from where they come. It is as if they appear out of nowhere, for one day there is no sign of them, and the next several thousand throng the plankton- and algae-rich waters. Supposedly the birds fly by night, the stars their guide as they travel thousands of kilometres across the African continent in this annual migration.

Both lesser and greater flamingoes occur here and freely intermingle. The

Flamingoes add vivid colour to the Pan.

birds prefer shallow, brackish or salty water and are often found in coastal lagoons and estuaries, thus the saline waters of a flooded Etosha Pan are ideally suited to their needs. Although similar in appearance, the two species occupy different niches in the feeding chain, the greater flamingo trampling the mud in shallow water and then wading with its bill inverted and filtering out small aquatic invertebrates, while the lesser flamingo walks or swims, swinging its head from side to side, also with its bill inverted and filtering food, mainly blue and green algae, from the surface.

Although the flamingo migration to Etosha and its subsidiary pans usually heralds the start of the annual breeding season, the birds are fastidious about their requirements and if enough rain has not fallen and the water is too shallow, they may abort the season's breeding. When conditions are suitable, however, the activities on the pan are fascinating. Both species build nests on cones of raised clay in shallow water, and usually lay one pale blue egg. The eggs are incubated by both sexes for about a month, and breeding is synchronized to ensure that all the chicks in the colony hatch about the same time. Once hatched thousands of chicks cluster together in communal nurseries, fed at first by the parents until they are able to forage in the shallow waters of the pan themselves. The initial struggle for survival is only beginning, and should the waters recede and evaporate before the chicks are fully mature, disaster threatens. The chicks then must trek across the caking mudflats, while their parents fly in relays to bring them life-sustaining food and water. The immutable laws of nature decree a regime of survival of the fittest, and here too only the strongest survive, truly earning their wings to eventually take their place among the departing flocks.

Some years ago the park rangers mounted an emergency programme which saw thousands of stricken chicks transported to water when the pan dried up before the birds' maturity. Subsequent proposals to build an artificial breeding platform, with its own water supply, to serve as a sanctuary in such times have, however, been rejected by park ecologists.

JULY 24. After spending the morning at Kalkheuwel where an elephant carcass lies, another victim of anthrax, we scan the bushes along Bloubokdraai, driving at walking pace as we look for leopards. Having spotted one here yesterday and two on the previous day, our hopes are high. Suddenly, there it is.

The leopard's yellow-green eyes fixed on something behind us ...

Although the leopard lies partially obscured by rocks and undergrowth, we decide to wait in the hope that it will move into better view. The sun sets without any sign of movement from the languid cat, until, as we're about to depart, it rises and slinks through the undergrowth, approaching to within about 15 metres of us before settling down again. In the failing light we can make out that it is a young male, with striking markings. Suddenly the leopard is alert, its piercing yellow-green eyes fixed on something behind us; turning we see a lone springbok walking past, which the leopard begins to stalk. Soon the cat is less than five metres from us, using our vehicle to screen it from the springbok, which continues walking steadily past, totally oblivious of the watching predator. But our hopes of seeing our first leopard kill in Etosha are dashed as the cat abandons the hunt, rises from where it's crouched and sits casually watching the antelope depart.

Springbok form the major prey of Etosha's leopards and the situation we have just witnessed was almost perfectly in the predator's favour, so the youngster's behaviour is inexplicable; even if not hungry leopard are known to hunt opportunistically, caching their prey in the branches of trees to feed on some time later. We can only put it down to bad luck, and perhaps lack of confidence on the part of the young leopard in the face of a fully grown territorial springbok ram.

JULY 27. Driving past Klein Namutoni waterhole we encounter a leopard about to cross the road in front of us. Apparently alarmed by the vehicle it stops in its tracks, then lowers itself to the ground and lies watching us intently. We immediately switch off the engine and sit very still, allowing the leopard time to realize that we pose no threat. It eventually blinks and shifts its gaze slightly, obviously more relaxed about our being there, and we use the opportunity to take some photographs. In time several other cars stop around us, reversing and manoeuvring for better vantage points. All the noise and activity scares the animal and it rises and moves into a dense thicket. More vehicles arrive and soon there are 15 of them creating a traffic jam, people shouting at each other from window to window. The leopard is now no more than a few tawny spots in the undergrowth, and there is little prospect of it moving while this commotion continues, so we leave, grateful for the 20 minutes we had had alone with it but annoyed at the human behaviour we have witnessed.

We drive slowly back to Klein Namutoni where we watch a giraffe 'gargle' with a bone it has picked up. The giraffe works the bone in and out of its mouth and appears to chew it with relish. It is the first time we have seen this unusual behaviour, known by the term 'pica' which means a craving for unsuitable food. Gemsbok, kudu and impala also are known to indulge in this practice in Etosha, which supposedly is indicative of a lack of calcium and phosphates in their diet.

Later the giraffe approaches the water where several others are already drinking, and is challenged by one of the males there. For the next 15 minutes we witness some determined neck-wrestling as the two stand side by side and crash their heavy heads and necks into each other's flanks.

AUGUST 3. We visit Ngobib fountain. There are 76 elephants trying to get to the water, which is accessible to only a few at a time as it lies in a deep hole among the rocks. The Etosha sun blazes down and elephant families cluster together in the little shade they can find, most of the trees being bare of leaves at this time of the year. There are a number of enormous bulls which have priority at the water, rumbling menacingly when any of the other elephants approach, and shoving aggressively if they fail to achieve the required reaction. One of the bulls has massive tusks – not long, but immensely thick. Etosha elephants generally have small, stunted tusks, often broken, as a result of the regional phosphate deficiency, although on one occasion we did see an elephant carrying tusks that would have gladdened the heart of a trophy hunter of old. Watching these animals now, it is difficult to understand the mentality that would encourage anyone to kill an elephant merely to mount some tusks on a wall, and perhaps to make a footstool from the feet; yet there is apparently no shortage of candidates in those few countries where elephant hunting is still legal.

Communication between elephants is complex, the animals using a highly developed infrasonic system of intense, low-frequency rumbles to 'talk' to each other over ranges of several kilometres. Until recently little was known about this, but research work done in Etosha has shown that this unique method of communication enables elephants to maintain contact with members of the herd and so maintain their elaborate hierarchical society, social

Two youngsters shove and push in half-serious play.

bonds that are essential to their survival in this remote and harsh environment.

Eventually it appears that all the elephants have had their turn at the water, for now groups begin to wander off. Several youngsters, teenagers perhaps, shove and push in half-serious play, unconsciously preparing themselves for the day when they too may seek dominance at a small waterhole, or need to keep competing males away from a cow in oestrus.

A young cow walks straight towards us, tentatively sniffing at the vehicle. Eventually she stands so close that she can run her trunk along the grille, cautiously testing the scent of this unusual object. After timeless moments she wheels away with a shrill blast through her trunk, flaps her ears in annoyance and turns to follow her group.

We spend the afternoon with a family of dik-diks. This antelope, weighing no more than about five kilograms, has striking similarities to the elephants, for it too has an elongated, prehensile snout reminiscent of a tiny trunk. Dik-diks also display strong social bonding, and clearly mark family territorial boundaries by smearing stalks of grass and bare twigs with a secretion from a gland in front of the eye.

AUGUST 11. We leave early to meet our pilot from Bushdrifters flying safaris, who has agreed to fly us over the pan for some aerial photography. After taking off from an airstrip just outside of the Park's boundary, we gain height and head for the pan. Now, for the first time we get a true impression of the vastness of the place. The pan itself stretches over the horizon, and even from a height of several thousand feet we cannot see the far sides. Looking down, we see a maze of game paths, well-worn tracks criss-crossing the barren surface and seemingly heading nowhere, from nowhere. Across to the west we see specks moving across the pan's surface and we change direction towards them. It is a group of zebras, walking steadfastly along one of the better-worn trails. As we swoop low for photographs they barely change pace, stare disconcertedly at our intrusion, then continue plodding southwards. They are kilometres from either shoreline, and are apparently trekking across the pan to its southern shores, where most of the zebras congregate at this time of year. Turning north again, we encounter more game, and zoom down for a better view of a few gemsbok which toss their long, straight horns in alarm

Dik-diks have elongated, prehensile snouts.

and gallop away from us, their hooves kicking up dust in the silvery early morning light. They look magnificent, the true creatures of the pan.

Flying closer to the northern shores now, in an area known as Poacher's Point, we see more animals, both on the pan itself and along its fringes. Wildebeest, zebras, springbok and giraffes ... but the one animal we are searching for eludes us. Although there are many elephant tracks crossing the pan we find none of the animals themselves, and turn south and east making our return towards Namutoni.

With the sun a little higher visibility has improved but still we see no more than an endless, ephemeral, pale expanse disappearing into a haze in the west. Our new perspective from the air truly justifies the one interpretation of the word 'Etosha' as 'the big white place'.

As we approach the eastern edge of the pan we notice large greenish patches on its surface. These are caused by the presence in the pan clays of

Several kudus wade into the water to quench their thirsts.

Flights of Namaqua sandgrouse come and go throughout the morning.

the mineral glauconite, a hydrate of potassium, iron and aluminium silicate. Flying back to the airstrip we pass over Namutoni and are given a better idea of the magnitude of running a game reserve – the complexities of management never seen by casual visitors – for sprawled out of sight around the gleaming tourist facilities are horse stables and workshops, vehicle parks, store-rooms, offices and staff accommodation, all essential components of the whole that is a national park.

AUGUST 13. Driving back to Okaukuejo we see that the country is beginning to show signs of grazing stress, for it is long since the last rains fell and the days are already hotter and longer. The few waterholes in the vicinity are well utilized, Gemsbokvlakte in particular having a never-ending procession of game drinking at it. Several thousand zebras throng the plains there, and barely a blade of grass remains for them to graze. At Okaukuejo the herds stand in battalions waiting to come to water, though they are so often prevented from drinking by lions in the vicinity.

AUGUST 15. 'The Old Girl' has been lying under the same tree for the past two days without making a kill, though she has tried unsuccessfully several times. Today there is tension in the air, for the animals must be parched. We set up our cameras early, ready for the kill we feel certain will happen. 'The Old Girl' is lean and hungry now, her last meal having been a small springbok three days ago, which she had to share with the rest of the pride.

Flights of Namaqua sandgrouse take off and land in relays throughout the early morning, and as the sun rises and the heat beats down, tension mounts. Springbok, herds of which have been shuffling around the waterhole all morning, begin to edge closer. They turn and flee, alerted by their own shadows, or by a low-flying bird, an eddy of wind and rustle of leaves, or some other real or imagined danger. 'The Old Girl' lies patiently in her shady lair, her head is motionless, but her ears are cocked and her eyes steadily fixed on the next meal.

A small group of springbok finally braves their way to the water. Several others follow, and then a family of kudu joins them, wading knee-deep into the water as they quench a thirst built up over the past few days. Through the foliage, we see the lioness shift her haunches, then launch herself from her hide-out. Straight and true she flies like a golden arrow; panic and confusion reigns as springbok and kudu explode from the water, and then barks of alarm rend the air. The lioness has chosen well: a young kudu trailing its parents flounders at the water's edge, then bolts headlong. But today the lioness makes no mistakes and with every bound, muscles bunching and stretching with each sinuous stride, gains ground and launches herself on to the frail young antelope as if in slow motion, bringing it to the ground in a crash of dust. Legs flail and a sad cry pierces the air; the other kudu stand 30 metres away, barking with agitation, dismayed, mournful.

The victim is in a stranglehold, the life-struggles dim, a weak kick, and then a chilling groan. A trickle of blood spills in the white chalk-dust. Around the waterhole, onlookers become voyeurs, having witnessed the kill so many visitors dream of seeing. Many turn away, embarrassed, and stunned by their intimacy with death. How many did not urge the little kudu to run faster, away, to safety?

At the waterside the dust settles, and 'the Old Girl' straddles her trophy and drags it to her lair. The other animals come to drink now, knowing that they are safe, one of their young having been sacrificed to the hungry predator.

The old lioness was a fixture at the Okaukuejo waterhole.

'The Old Girl' was a familiar figure at the Okaukuejo waterhole, and for many visitors she was their first-ever lion, perhaps for many, their only lion. She took up residence at the waterhole, probably too old to roam the large territory held by her pride, and honed her hunting technique so she was successful in about one in three attempts – exceptionally high for a lion, particularly an aged, lone lioness. For months 'the Old Girl' stayed at Okaukuejo, often alone, occasionally with the other members of her pride.

The kill we witnessed here was one of the last occasions on which we saw 'the Old Girl', and some time later we were told by park rangers that she had been killed by a farmer a short way across the southern boundary. Several other members of her pride were also poisoned or shot by the same person.

Lions, like all major predators, are anathema to stock farmers and are shot on sight by most. They are also almost impossible to fence in, for a lion will climb, bite or dig its way out, through or under the wire which is in the way of its will. Electrified fencing is effective, but is expensive to install and maintain. Until the funding is found to erect such a fence around Etosha, farmers will continue killing lions, and other animals that stray on to their lands. Some say these animals are even enticed there, with baits staked out to attract them, and giving someone occasion to boost their ego with the claim: 'I shot a lion'.

Lions are not endangered in Africa. Not yet. However, there were also 100 000 black rhinos roaming the continent a mere 30 years ago, which then were not endangered either. Lion probably rates as the single most important species in terms of tourist value, and without tourists, game reserves in most of Africa could not survive. It follows then that everything should be done by the authorities to protect their lion populations. Perhaps the first step should be to remove the loopholes in the laws which allow farmers to destroy a part of Africa's natural heritage. Farmers on lands adjacent to game reserves welcome the benefits they derive from their proximity; so too should they accept the difficulties.

Have we not yet learnt from the rhino saga, or will we wait until lions are on the brink of disaster, set to disappear from all Africa's wild places, to become mere zoo showpieces?

THE WET & THE DRY

'There are two Africas and I do not know which I love the best: the green,
lush, bright country when the sap is running and the earth is wet; or the dry,
brown-gold wastes of the drought, when the sky closes down, hazy and
smoke-dimmed, and the sun is copper-coloured and distorted.'
Doris Lessing, *Going Home*

Goas fountain in the wet and green summer months ... and in the harsh dry winter.

Etosha's year can be divided into two distinct seasons, the wet and the dry, each with an entirely different character and each an entirely different experience.

The rainy season begins in about November and continues through March and, in good years, into April, with peaks usually in late January and February. The average precipitation over the Park ranges between 400 and 450 mm. This is also the green season, when the Park blooms into a floral spectacular and the vegetation grows lush and verdant. It is a time of birds and babies, for these are the summer months when most of the animals drop their young, and birds flock from across the continent. In years of good rain the flamingoes number in the millions, turning the partially flooded pan into an endless vista of pinks and whites. Etosha has more than 340 bird species, and it is during these summer months when the majority of them can be seen.

Summer is not, however, the prime game-viewing season. Hot and humid, the average daily maximum for December is 35 °C; rains fill the usually dry pans and depressions throughout the Park and the game disperses, no longer dependent upon a limited number of drinking places. The massive herds that throng the plains around Halali and Gemsbokvlakte in the dry winter months migrate to the wetter western areas where they can utilize grazing inaccessible to them before. Most of the predators follow in the footsteps of their prey, until grasslands that throbbed with life only a few months earlier stand deserted. Many of the elephants leave the Park altogether, trekking across its northern and western boundaries in movements still not entirely understood.

The cool, dry winter is therefore the best time for visitors intent on seeing large numbers of big game, for with all dependent on the limited number of waterpoints, wildlife congregations are nothing short of spectacular. Predators never move far from the water, and trees and shrubs shed their leaves so cover is minimal and viewing much easier.

Winters are cold, particularly in the early mornings, with the average minimum for July being as low as 6 °C, although the days are sunny and mild. However, Etosha being in the tropics (latitude 19 °S) winters are short-lived and by mid-August daily temperatures are already beginning to climb. In fact the dry season can be divided into cold-and-dry and hot-and-dry months, the latter a time of shimmering heat hazes with long lines of animals trekking dazedly across the searing plains to seek relief at one of the waterholes.

This is a time of gusting dust storms, winds that whip across the open plains, and evenings when the western horizon is burnished a fiery red as the setting sun ignites the haze. Towering whirlwinds, laden with dust and debris, spiral across the pan and soar spontaneously into the glaring, pale blue sky, while plodding wildebeest and zebra, like ghostly apparitions, emerge from out of the swirling obscurity.

SEPTEMBER 2. We returned to Halali from Okaukuejo yesterday. Driving here we passed massive herds of zebra and wildebeest on the plains around Charitsaub and Salvadora, more than we have ever seen before. The game seems to be heavily concentrated in this area now – springbok, gemsbok, wildebeest and zebras in particular, as well as a great many elephants. On the circular drive near Rietfontein we encountered a pair of secretary birds on a nest alongside the track, which performed for us in a form of courtship ritual we have not previously seen. An added bonus soon after leaving the birds was the sighting of a big rhino bull making his way down to a seepage near the edge of the pan.

Spring permeates the air at Halali, and the birdlife is becoming more colourful although the countryside is dry and the midday heat oppressive.

We left camp early this morning and drove to Charitsaub, where we have encountered 11 lions, part of the well-known Charitsaub pride. There are seven cubs, now almost too big to be so called, three lionesses and the magnificent gold-and-black-maned pride male. These lions probably form part of a much larger pride when they join up with a sub-group we know as the 'Rietfontein lions', comprising a young male, two lionesses and five small cubs. We believe the two groups have split temporarily while the lionesses raise the variously sized cubs. The two groups use the same territory, and we have seen the pride male and the oldest lioness with both groups.

Many of Etosha's lions have been branded on their flanks or shoulders to aid researchers in tracking them. Though unobtrusive, these marks have been useful to us in identifying lions we have seen and photographed before, particularly when there are few other identifying features to guide us. However, we have found that most lions show readily recognizable behaviour traits, and after spending long periods with certain prides we can quickly identify individuals by their 'personalities'. Even the cubs show their characters at an early age, some apparently far braver than others. One young sub-adult male in this group is obsessed with our vehicle, and regularly breaks off whatever other activity he's engaged in to stalk us, losing courage only when he is about two metres away.

The presence of the lions means that the other game cannot come down to drink, herds mingling and jostling in the heat all around us now. From time to time the lions sit up where they're lying in the grass to watch these animals, but they show no interest other than that. We leave the lions, certain they will do nothing until the coolness of the late afternoon, and drive through the impatient herds to Rietfontein. As we enter the mopane woodland that encircles Rietfontein waterhole, we slow down to a walking pace, knowing that two leopards are regularly seen along this stretch of road. Our diligence is rewarded, for we find a leopard lounging atop a fallen tree-trunk, in full view. We watch, infatuated with this regal cat, until a second leopard appears, walking to the first and greeting it with affection. The second leopard is quite a bit larger, and on seeing that it is a female we deduce that it is the mother. They relax together for a while, and then the young male rises, stretches, and walks into the undergrowth and begins feeding on something in the grass. Only now do we realize that the leopards have a kill here, although we can barely discern the cat's outline as it tears at the carcass.

The leopards show no signs of moving before nightfall, so we decide to leave them for a while, but return later to find them still resting beside the dead tree where they have become a major tourist attraction.

SEPTEMBER 5. Rising with the first weak light of morning, we leave Halali and head west intending to 'do' the circular drive around Rietfontein, a prosperous area for us in the past few days. Shortly after leaving camp we catch sight of a hyaena loping down the track, and immediately hereafter see three cheetahs which cross the road and walk into a thorn thicket. We try following them, but eventually the bush becomes too dense and we lose sight of them.

Half an hour later we find the Charitsaub lions, although today the big male is not with them. Our attention was drawn to them by a herd of zebras, their snorts of alarm having brought all the animals on the open plains to attention, now all staring intently to where the lions lie in knee-high grass. The goings-on are amusing to watch, for every time a lioness begins a stealthy stalk, her efforts to remain concealed are foiled by a curious cub, eager to see what the adults are up to, and sitting up in full view of the intended victims. After a fruitless hour the lions give up and begin walking towards Etosha Pan which lies a few kilometres to the north. We follow them, enjoying the attentions of three of the cubs which lie in ambush for our vehicle, jumping out at the last moment when we approach, then rushing to safety alongside the adults. Eventually we reach the pan, glaring white under the morning sun, mirages already shimmering across its surface. The lions lead us to a seepage we have not seen before, just a few small pools of water almost hidden beneath scattered tufts of *Sporobolus spicatus*. After quenching their thirst, the lions seek shade among the grassy tufts, and within ten minutes are asleep. We decide to wait, for there are signs that other animals also use this contact spring, and the attempts at stalking earlier could mean that the lions are hungry.

Wildlife photography frequently entails hours of inactivity, so we are always well stocked with food and refreshments, a small gas cooker and water. On occasion we have to stay out all day and night, waiting for something to happen, so we are well prepared with a selection of reading matter too. Now, we sit patiently reading, trying to get comfortable amid camera equipment, binoculars and all the associated paraphernalia of our profession.

The two 'Rietfontein' leopards, mother and child.

The lions begin walking towards Etosha Pan, a few kilometres to the north.

Sometimes patience pays dividends – and after about three hours we see several groups of game converging on the spring; at least 40 springbok, a dozen or more zebras and a handful of wildebeest. We reposition our vehicle for the best vantage, select lenses and ensure all cameras have fresh rolls of film. The lions, too, choose their positions, and we watch with mounting anticipation as one lioness stealthily moves around in an outflanking manoeuvre, the other two slinking into an ambush behind tufts of grass. The cubs also appear to take the impending hunt seriously, and hide themselves amid clumps of grass, although from time to time a curious head rises to peer about.

Now the animals are standing nervously looking at the water, apparently having sensed or smelt the presence of lions, for they toss their heads and snort anxiously, stamp their feet and circle about, on edge. Suddenly a streak of gold is flying towards the nearest zebra, bounding across the intervening ground with unbelievable speed; but the alarm has been raised, the attack begun too soon, and amid clouds of dust the zebras, wildebeest and springbok gallop to safety. The lioness chases for about 200 metres, but she is losing ground all the time and soon breaks off and walks back, panting heavily, to where the others now sit watching. She drinks, and the lionesses again set up their ambush, but after almost an hour's wait it is obvious that the other animals have no intentions of returning.

The rest of the afternoon passes uneventfully, and we leave the lions before sunset to drive back to our camp, passing several hyaenas and a pair of Cape foxes – among the rarest of predators in the Park. As the sun dips below the horizon we see a black rhino in the thorn-scrub, and on closer investigation discover he has three horns. Although there are records of two rhinos in Etosha with a third horn, we are taken by surprise as this is an extremely rare condition, and previous sightings have been in other areas. Moreover, unlike the other two rhinos which are reported to have knobby lumps as the extra horn, this bull has a perfectly formed and shaped horn, higher up on the forehead than the normal posterior horn.

SEPTEMBER 20. We set off before sunrise for Goas, an artesian spring about 18 kilometres from Halali where friends have reported seeing four cheetahs. Arriving just as the sun lights the treetops, we watch dawn roll over the countryside. The waterhole is disappointingly deserted, but we find a diversion in some whitebrowed sparrowweavers making home improvements at a nearby nest. Suddenly we notice movement on the far side of the waterhole, and realize that it's a group of cats. Thinking we have found the cheetahs we are looking for, we quickly raise binoculars – to three leopards loping across the road: an adult female and her two half-grown cubs.

We leave Goas once the leopards have disappeared into thick bush, and visit two other springs in the area without success. Then to our surprise we see another leopard 'stalking' a group of giraffes near the turn-off to Kamaseb. The giraffes stand peering over the treetops at the cat – a young sub-adult male. It's an amusing sight, for the giraffes are of course far too big for a leopard to consider, and they appear somewhat bemused by his approach. However, in time they canter away through the mopane with their typical rocking-horse gait, and the leopard gives up in disgust.

After stopping for a picnic lunch under a shady tree, we drive back to Rietfontein, one of our favourite waterholes, and well used by lions, leopards, cheetahs, rhinos and elephants, as well as by all the herds of herbivores in the area. Driving slowly to the parking lot, we see two lions, a big black-maned male and an old lioness, lying on a slight rise overlooking the water. Then from behind we see the Rietfontein 'pride' – a young male, two lionesses and five young cubs – approaching the water. The other two lions don't pay much attention to the new arrivals and we recognize them as members of the nearby Charitsaub pride. After a while the old lioness joins the larger group where they are now lying, and the mutual greetings and subsequent grooming of each other reinforce our beliefs that these prides are sub-groups of a larger one. We note with interest that the male usually encountered with these cubs, a young adult only beginning to grow his mane, remains wary of the black-maned male which lies some distance away.

The cubs engage in their usual waterhole antics, jumping on and over each other, 'attacking' the young male lion which responds with tolerance and seems to enjoy playing with them, and stalking a few red-billed teals at the waterside. Suddenly the cubs stop their playing, and the adults sit up where they're lying, all staring into the mopane trees that grow on the southern side of the waterhole. We look but see nothing, and hear nothing. Five minutes pass before we see a herd of elephants appear, breaking into a shambling run when they clear the trees and see the water. They are obviously hot and dusty, for the entire herd wades deep into the water, slurping and splashing with abandon. The lions flop down to sleep again, though the cubs sit watching the elephants in fascination for some time before resuming their games.

It is after sundown, the sky a pinkish purple, when a black rhino appears out of the scrub and approaches the water. Although apprehensive, he ventures to the water and drinks. We sit in awe at the sights around us; considering too the leopards of this morning. This has been an exceptional day.

SEPTEMBER 24. We spend the day scouring the plains around Halali, still searching for the elusive cheetahs, but with little success. We do see lions, eight at Rietfontein in the morning, 11 near Salvadora, and later four at Helio, the waterhole right near Halali. The annoying thing is that the cheetahs are here, and have been seen regularly by other visitors.

SEPTEMBER 25. We decide to break from our cheetah search, and drive to Agab where we plan to spend the day at the waterhole. There is a lot more game here than on our last visit, and we spend a profitable morning filming zebras,

A melanistic Gabar goshawk makes a meal of an agama lizard it has caught.

The lion pride rests after its meal.

The rock monitor, or leguaan, has the highest biomass of all predators in Etosha.

hartebeest and birds, including a bateleur eagle which comes to drink right alongside us. Several giraffes arrive and after warily watching us for about an hour, approach the water. Later, on our drive 'home', we encounter the seven Goas pride lions walking along the road. As we drive up they lie down in our path, rolling in the dusty track and staring balefully at the vehicle.

SEPTEMBER 26. After an uneventful morning we find the Rietfontein lions with a springbok they have just killed. The animal is in the final throes of death, kicking in the jaws of the young male which appears to have made the kill. The rest of the pride is soon milling around, squabbling, growling and ripping into the carcass in a feeding frenzy; we fear for the lives of the young cubs which are in the midst of the orgy, trying to get their share of what will be little more than a snack for the three adults. The young male grabs a corner of the carcass and tugs at it, trying to drag it away from the others. Immediately the two lionesses bite hold, and a tug of war ensues, the cubs dangling from the prize in between, gobbling and swallowing at every opportunity. The adults lash out at each other fiercely, snapping and swiping with full-blooded blows as they struggle to get more than a fair share.

The meal lasts less than four minutes until the male manages to tear off a large portion, including the head which swings grotesquely as he runs off, one cub hanging on for dear life. The two lionesses and the other four cubs continue pulling at what remains. Suddenly one lioness rounds on the other and batters her to the ground, standing snarling over her with teeth bared. Cowering, the lioness slinks away, and the victor stands back to give the cubs the remaining spoils, even tearing pieces of flesh from the remaining bones and feeding them to her youngsters which settle down to gnaw at the last bones, hissing and spitting at each other occasionally.

SEPTEMBER 27. Though it's Daryl's birthday we resist the temptation to laziness and drive out towards Charitsaub and Salvadora, again in search of the cheetahs. In more than four months we have had very few good sightings of these cats. Soon after sunrise we leave camp, to a glorious spring morning.

Less than an hour later we find what we are looking for: a lone cheetah has killed a springbok near the junction of the main road and the Rietfontein 'ompad'. Although we missed the actual kill, she has just started eating, and the victim has barely a mark on it. She feeds greedily, starting near the rump and working her way forward, glancing our way from time to time but apparently more concerned about the possibility of some other, more powerful predator arriving on the scene and stealing her meal. Soon her jowls are caked red, and her stomach swells visibly as we watch her eat. Her victim is a fully grown springbok male, almost 50 kilograms in weight, and she is doing her best to devour it all. The feast continues for more than an hour, during which time the cheetah barely stops eating, other than to look around. Thinking back on the lion feasts we have watched, the contrast in sound is striking, the bloodcurdling growls and snarls of the lions replaced by an eerie silence here, only the occasional crunch of a bone breaking the quiet.

Eventually little more than the head, neck, bones and a blood-spattered skin remain. The cheetah walks ponderously, her belly swaying and hanging heavily, into a dense copse of thorn trees nearby where she lies grooming herself. Happy with our luck, we declare the rest of the day a holiday and return to camp for brunch.

The cheetah feeds greedily, starting near the rump and working forward.

THE ONSET OF THE RAINS

Summer comes early in Etosha, with temperatures rising noticeably after the advent of spring. It is a time of blinding heat, the mopane trees bare of leaves and bereft of shade and shelter, and waterholes shrinking until the animals are forced to queue at the brackish, tepid puddles.

Every day now, eyes turn skywards, as occasional clouds scud across the aching heavens only to swirl and dissolve over the uncertain, dancing horizons. Fierce and unremitting, the sun continues its relentless assault, draining life from whatever it touches. But with every day the billowing masses offer palpable promise, and expectancy mounts. Occasional cold spells, rolling in from the Atlantic coast in the west, bring respite but never for long. Lines of wildebeest trek forlornly across the parched plains, and zebras stand bemused in the glaring heat.

October offers hope, and as the days pass the clouds build, dark and reassuring, heavy with rain. Although the rainy season is really from January to March, 'short rains' can be expected this month, which may come as light showers or as heavy downpours.

OCTOBER 7. Driving back to Halali from Okaukuejo where we have been to collect our mail, we watch black thunderclouds building in the late afternoon sky. Dramatic bolts of lightning crackle across the heavens, and static charges the dry air. Dust-devils swirl spontaneously across the road, and we see a monitor lizard on the move – a sure sign of impending rain.

Breaking out of the mopane and thorn scrub into open country, we watch a purple-black rainstorm blowing across the immense Etosha landscape, sudden squalls of rain bringing renewal to the desiccated countryside. We find the Rietfontein lions, soaked through but revelling in the wetness, a welcome relief from the remorseless heat of the past weeks. Springbok, which just hours earlier stood listlessly nibbling at dehydrated acacia shoots and dry grasses, 'pronk' exuberantly across the landscape, bouncing high, feet together, in stiff-legged stots which almost impel one to join them.

Within minutes the roadside runs with gurgling rivulets, dry-caked hoof-marks fill with water and parched mopane trees shake off their lethargy and glisten with renewed life. Then, as quickly as it arrived, the storm passes, rolling thunder echoes in the distance across the steaming pan, and a dazzling rainbow enriches the glittering land. We drive back to camp in the last sweet light of evening, splashing through puddles and over sheets of water that have turned the tracks into hazardous slipways.

OCTOBER 10. It has rained all night, heavy drops that burst on our tent like swollen teardrops, and we waken to an enchanted world, sodden trees sparkling as though bedecked for an early Christmas. Thunder still echoes in the distance as we set off for Rietfontein, hoping to find the lions. Research has shown these predators to be particularly active and able hunters during thunderstorms, for the noise conceals any sounds they themselves may make in approaching their prey.

We pass many herds of game along the way, moving towards where the heaviest rains have fallen. If the rains continue, the animals will have departed from this area within days, trekking across the Park to the summer grazing grounds in the mid-west. Already a restlessness is detectable among them and herds, previously scattered across the vast open plains, begin massing.

Passing Rietfontein, which we find quiet, we turn down the southern leg of the circular drive in time to see the Charitsaub pride pulling a springbok down in an open area near a belt of thornbush. The rest of the herd is fleeing wildly with three of the lion cubs in pursuit, but the cubs give up their chase and return to where the pride is beginning to feed. It seems that the kill was an opportunistic one, occurring only because the springbok happened to pass within range and not because the lions were hungry, for although the cubs and three lionesses feed on it, the two big males show total disinterest and lie apart some distance away.

No more than a kilometre away, we find the Rietfontein lions, also feeding on a springbok, though very little remains. The elder lioness from the Charitsaub group arrives and greets the two Rietfontein lionesses, rubbing faces with them. We then hear one of the Charitsaub lions roaring in the distance – from our position we can actually see both prides – and soon the Rietfontein pride begins walking in that direction.

We take the road back to the Charitsaub group, the members having moved some distance away from the two males and now watching the other group walk towards them. The Rietfontein group changes course slightly and passes within about 150 metres of the larger pride, neither group acknowledging the presence of the other. We watch, fascinated, as the Rietfontein pride now joins the two males whom we had seen earlier with the Charitsaub pride – once again reinforcing our belief that the two 'cub groups' are actually constituents of a larger territorial pride.

Thunderclouds build up in the late afternoon sky, and squalls bring temporary relief.

The rest of the morning passes amid occasional rain showers and gloomy, overcast weather, until the sun breaks through late in the afternoon. Sitting at Rietfontein waterhole we see a melanistic Gabar goshawk swoop down and catch an agama lizard sunning itself on a rock. It perches on a tree alongside us and proceeds to tear its prey to pieces, feeding voraciously.

OCTOBER 13. It is our last day in the Halali area, for tomorrow we must return to Okaukuejo and prepare for a trip that will take us away from Etosha for almost two months. We are saddened at the thought but know that when we return the rains will be upon us and Etosha will be transformed for a brief season into lush greenness.

The clouds have disappeared once more and the sun beats down with the force of a furnace. The rains of the past few days appear to have had little effect other than to raise humidity levels, and there is still very little water in the pans and waterholes. Shortly after sun-up we drive across the Halali plains and take the northern loop road before entering the mopane woodland near Rietfontein – a farewell bid to our many animal acquaintances. We pass a lone lion on a zebra carcass – an anthrax mortality – and not much further away find three cheetahs sitting beneath an isolated tree in the middle of the grasslands. Now and again one of them climbs into the lower branches of the tree and peers out across the plain, but there is little game in the vicinity and eventually they all flop down gracefully in the shade.

A lone springbok appears and the cheetahs assume attack positions, loose-limbed as they crouch in the grass. But the springbok veers away. They get up from where they lie and prowl about the long grass, flushing a scrub hare and giving chase. Tails flailing like rudders in the air, the three cats gain on the unfortunate hare until with a final graceful bound it is snatched into the jaws of the largest of the trio, and vigorously shaken. Head held high, the cheetah trots proudly away with its spoils while the other two watch hopefully, but make no attempt to take the meal from the victor.

Later, a movement in the thick thorn scrub gives way to a rhino, heading in our direction. As it breaks into the open we identify it as the three-horned animal we had previously seen in bad light. The third horn is clearly visible, about 10 centimetres long and perfectly shaped. According to our queries with park rangers he is not an animal they have identified before, so we are doubly excited at our discovery. He walks obliquely towards us, stopping occasionally to test the wind, peering short-sightedly about but apparently unaware of our presence. He passes within 15 metres of us, and walks calmly into another thicket similar to the one he'd come from, browsing as he moves.

We spend the late afternoon at Koinseb, an artesian spring deep in the mopane woodland off Elandsdraai, and are about to leave after an uneventful couple of hours when we notice some springbok, which had been grazing quietly on the green grass around the water, whistling in alarm and staring towards the trees at the western side of the spring. We study the area and see a leopard moving gracefully through the trees, but he ignores the springbok, and they return to their feeding, keeping a wary eye on the predator all the same. The leopard pays us no notice and approaches to within about 25 metres of our 4 × 4 vehicle. He stands and watches us over his shoulder before crouching to drink deeply. We stay until night falls and then drive back in the dark, watching nightjars play in the twin beams of light ahead of us.

DECEMBER 18. Driving through the Andersson Gate, we arrive back in Etosha National Park after an absence of approximately eight weeks to find the land parched and withered under a baking sun. Apart from a few isolated cloudbursts the summer rains have held off, and though the trees are green and lush the veld is now brown and dust-laden. Fires, started by park rangers as part of the veld management strategy, as well as by lightning strikes, have swept across vast expanses of the Park, giving the landscape a harsh and sombre appearance.

The Okaukuejo region has had the least rain of all areas, so we head for Halali where we find the first signs of a green flush in the burned areas, and more game. Springbok and zebras appear to be gathering in readiness on the plains for their westward trek, and many of the females are in advanced stages of pregnancy. Although we will base ourselves in Halali for the summer months, we continue to Namutoni to reacquaint ourselves fully with the Park, and also because the eastern sector usually records the greatest volume of early rainfall.

Along the way there are encouraging signs of recent falls; in places muddy puddles line the road, and the grass is greener. We also find higher concentrations of game, and are astonished at the number of giraffes in the area. Fischer's Pan is filling, and the first flights of flamingoes have arrived, several hundred of these beautiful pink and white birds bringing vivid colour to the drab mud flats.

We follow the route around Fischer's Pan and turn in at Aroe waterhole to find a large herd of elephants quenching their thirst. They drink and splash for some time, the babies revelling as usual in the water and scampering around, through the legs and under the bellies of the adults.

On the road back towards Namutoni we notice far greater numbers of gemsbok than we have seen in previous months, in addition to large herds of wildebeest, zebras and springbok. The tension in the air is almost tangible, the animals restless as if they too are urging the rains to fall so that the migrations can begin.

The system of rotational grazing that results from these annual game movements, although now seriously restricted by the erection of the northern boundary fence, is essential to the well-being of the grasslands, and experts maintain that the future of this national park would be better served by re-aligning the fences to include the traditional summer grazing lands around Lake Oponono in Ovamboland to the far north.

DECEMBER 20. Although clouds hang in the sky at dawn, the morning heat soon burns them up and the clear blue heavens give no promise of respite. We tour Bloubokdraai near Namutoni, but the summer vegetation is far denser than before and we see few of the Damara dik-dik we know inhabit the area. Later in the morning we capture a harmless sand snake near our tent in the campsite, and release it out of harm's way in some thick bush.

In the afternoon we travel back to Halali, seeing little game on the way until we encounter a family of bat-eared foxes near Goas. We watch them, two adults and their four offspring, and note the location of their den in some burrows on the edge of a clearing. The foxes are very shy, the pups quickly disappearing down the burrow when we approach. However, both adults lie in the shade of an acacia thorn nearby, at first watching us cautiously, but eventually regarding us with less suspicion. After some time two of the pups raise their heads out of the burrow, but soon disappear again. We leave the family, deciding that we will habituate them gradually to our presence.

Our 'Christmas' leopard, resplendent amid emerald-green summer growth.

DECEMBER 25. After a late breakfast we are on the road to Namutoni, stopping first at the foxes' den. It is clearly too late in the morning, however, and we don't see any of them. Bat-eared foxes are primarily nocturnal, and are usually only seen by day in the very early mornings or late evenings. We continue to Goas, stopping to look at a group of lanner falcons perched in a dead tree, but then notice a leopard slinking through the shadows. It continues in our direction, crosses the road only metres from the vehicle, then doubles back and sits scrutinizing us at close range before moving off to lie in the thick bush.

Continuing towards Namutoni, we are surprised to encounter a herd of some 50 elephants feeding in the mopane, having been led to understand that all these animals move from this section of the Park to their summer feeding grounds in the south and north-west. At Kalkheuwel waterhole a group of about 40 kudu is gathered, along with a lone African spoonbill which wades up and down, feeding in the shallow, muddy waters.

Later at Chudop, we find more than 50 giraffes gathered about the spring. They take turns at drinking and several young males engage in sparring and neck-wrestling tussles, their heavy necks and heads swinging and crashing into each other in playful emulation of the battles adult bulls enter into over females in oestrous. This behaviour is also said to lead to cohesion within herd groups. While it is generally believed that giraffes do not defend territories, we see a large, dark bull deliberately drive off another bull which comes to drink, approaching from a direction different to this group.

The blazing sun soon drives us away from the waterhole and we make for Namutoni, where we can relax under shady trees until the cooler hours of the late afternoon. Summer temperatures in Etosha can soar well above 40 °C, and even the animals seek shelter from the midday heat.

On our way back to Halali before sunset, we encounter two lions strolling along the road – the first we have seen since our return to the Park a week ago, and a fitting close to our Christmas day in the bush.

DECEMBER 26. After an uneventful morning we return to Goas waterhole in the late afternoon, where a young leopard is drinking at a pool by the roadside. The leopard watches attentively while we manoeuvre the vehicle into position, but relaxes and resumes drinking once the motor is switched off. The big cat drinks continuously for about 10 minutes before retiring to a patch of shade. Shortly, a group of springbok passes nearby, and the leopard rises to a crouch in readiness. None of them come close enough, however, and the leopard sits watching as the springbok continue on their way. Perhaps it is still too hot to embark on a chase, for soon the cat returns to the water to drink.

While we are watching the leopard a family of shrikes flies down to the water, veering off with raucous alarm calls at the last minute, when they spot the cat lying nearby. They settle in the treetop above the resting leopard, and display their annoyance with a cacophony of irate chatter. The leopard is visibly irritated, for it raises its head and snarls aggressively, lips curled back and whiskers twitching. We stay with the leopard for almost two hours, but shortly before 19h00 it rises and walks into the bush, concluding our time with it.

DECEMBER 27. We find the Rietfontein lion pride at their regular haunt – the waterhole after which we have named them – and are pleased to see that all five cubs are still alive and well. Mortality rates are generally high among lion cubs, as much as 60 per cent in Etosha, but these cubs have clearly survived the dry season and still appear to be fit and healthy despite the reduced concentrations of game – representing potential food. The rainy summer season will be the vital test, for hunting is far more demanding for these predators at this time. Lion cubs depend upon their mothers and other pride members for food until their second year when they begin to hunt for themselves, and starvation is a major cause of cub mortality.

The rains are all around us now, dark clouds blot out the sun and gusting squalls add to an ominous quality over the land. But in our vicinity only a few heavy drops fall, throwing up dust where they hit the parched earth. Across the plains come the curious honking barks of zebras excited by the prospect of rain, and everywhere kori bustards strut in magnificent displays, their tail and throat feathers puffed out and extended and their deep, booming calls resonating through the countryside.

DECEMBER 30. It has rained heavily in the afternoons of the past two days, and now the land sparkles, renewed. Though still too soon for new growth to

The giraffes take turns at drinking, legs widely splayed.

have made an appearance, the pools of water that dot the landscape have already attracted migrant waterbirds; stilts, sandpipers, avocets, plovers and several varieties of waterfowl collecting where previously there were only dusty plains. The vegetation, washed clean of its mantle of dust, glistens with life and the air has a fresh, clean scent to it.

Leaving Halali before sunrise, we reach the bat-eared fox den as the first rays of sun catch the treetops. The female and her four pups are already lying on a mound of sand outside their burrow, huddled together before the gently warming rays of the sun. The family has become accustomed to us now, for we have paid them several visits since first encountering them, and they engage in their activities with barely a glance in our direction.

As the sun rises higher it casts its light over the fox family, enabling us to photograph them as they go about their grooming. Two of the pups wander about, foraging for insects under stones and pieces of dead wood, their big ears rotating like radar as they listen for any give-away sounds. Finally they retire to their burrows to sleep through the heat of the day.

Continuing on our way we see our first new-born springbok lamb, perhaps only an hour or two old, and later we pass a herd of hartebeest with several young. Now that the rains have started the calving should begin in earnest, and we expect to see more and more young in the ensuing weeks.

At Agab we find two male lions and a lioness lying under some trees. It appears that the larger of the males is courting the lioness, but we see no action in the time we spend with them, and decide to head back to camp.

In the afternoon, en route to Rietfontein, we encounter a rhino but cannot get a clear view of it before it runs into a thicket, and then just past the waterhole find the young male leopard we have seen frequently in this area. He is lying in a depression alongside the road chewing on a terrapin that he has probably caught in the rainwater pool, an unusual snack which he seems to relish. Soon it begins to rain, huge, heavy drops which fall with increasing tempo, beating a tattoo on the roof of the vehicle and obscuring our vision. The leopard takes no notice of the rain and continues with his feeding even though his coat is soon soaked through.

Finally the leopard finishes his meal and moves deeper into the bush, a signal for us to return to camp.

The bat-eared fox family lies huddled before the gently warming rays of the sun.

The lions sit grooming themselves while a brilliant rainbow appears in the distance.

JANUARY 3. Thunder rolls across the heavens and dramatic purple-black rainclouds mass above us. Great bolts of lightning smash into the earth, and a gusting north-easterly wind whips the vegetation in fury. We sit amid a mounting storm at Salvadora, the seepage on the edge of Etosha Pan near Halali, while ten lions, the seven sub-adults and three lionesses of the Charitsaub pride, shelter nearby.

Soon the storm is upon us, and the lions bunch together, tails to the driven squalls, and, perhaps through some instinctive knowledge, having moved out from under the tree where they were lying. Of all the big cats, lions appear most forlorn and bedraggled in the rain, and these before us are a sorry sight as they huddle together against the conditions. The downpour does not last long, however, although the countryside runs with water and huge puddles lie everywhere. These torrential cloudbursts, typical of the early part of the rainy season, are of less benefit than the long, soaking showers which come later, for now most of the water is lost in run-off before the hard, dry earth can absorb it. Because the storms are short-lived, the ensuing hot, sunny weather serves only to evaporate most of the rain fallen.

The lions sit grooming themselves, and each other, and some of the sub-adults leap about in high spirits. A brilliant rainbow appears in the distance, and serene silence settles over the land.

JANUARY 5. Surprisingly, the animals show little joy at the arrival of the rains. Springbok stand in cheerless groups, hunched against the driving rain as they patiently wait out another storm. One of them lies nonchalantly, as if resigned to being drenched through, in a growing pool of water. Abdim's storks, which appeared in their hundreds as if from nowhere with the onset of the rains, strut in serried ranks through the dripping undergrowth, sleek and glossy-black against the soaking land.

This is the heaviest rainfall we have had this summer: more than 40 millimetres falls in less than an hour. Sitting at Rietfontein, we watch trickles become rivulets, gurgling down game trails and rushing into catchments that soon fill and overflow. The dry dust landscape is transformed.

The summer rains are a major climatic phenomenon in Etosha. Although by tropical standards the precipitation is not high, ranging from an annual 500 millimetres in the eastern sector of the Park to 400 millimetres in the central regions and a low 250 millimetres in the arid west around Otjovasandu, it is a revolutionizing force here, governing the tempo of life in all forms.

While the wet land steams under the heat of the African sun, still penetrating the dark, towering cumulonimbus clouds, many of the herbivores begin their migrations to the wet-season grazing lands, and numerous bird species begin nest-building and courtships that frequently entail spectacular displays and dramatic territorial disputes. Most prominent, perhaps, are the black korhaans, their harsh 'kraak-a-kraak-a-kraak' calls echoing across the plains while their dare-devil relatives, the redcrested korhaans, perform kamikaze aerial feats, soaring skywards, then folding their wings, collapsing, and plunging earthwards as if shot in mid-flight, only to open their wings and level out scant centimetres above the ground.

Now as the storm moves on, the western sky begins to glow with promise. We drive to a vantage point on the edge of Etosha Pan – a surreal, glittering waterscape where scant hours ago lay stark, featureless salt-sand – and watch the colourful display as day surrenders to night.

JANUARY 7. We returned to the Namutoni area yesterday after having heard that both the springbok and wildebeest have begun dropping their young. Now, driving along the Fischer's Pan-Aroe road we see ample evidence of this ourselves in the many young animals among the herds. It is a cool, heavily overcast day and rain showers fall from time to time – conditions which stimulate parturition. We stop and with binoculars study each and every springbok ewe and wildebeest cow, hoping to find one in the initial throes of labour so that we may photograph the birth.

Eventually we find a wildebeest lying in the grass on the edge of the pan, a sticky, wet, and weak calf stretched out behind her. Almost immediately the cow rises to her feet, the placenta still attached to her, and licks her newborn roughly, quickly eating the clinging foetal membrane. Within minutes the calf rises shakily on to unsteady, spindly legs, and follows its mother to join the herd which has waited nearby throughout this event. Several herd members gather to inspect the new addition to the group. Then all walk further out on to the open pan, perhaps to afford them greater protection because of a lack of cover there for lurking predators.

Wildebeest young are born after a gestation period of about 250 days, and have a mass of between 20 and 23 kilograms at birth. In Etosha the majority of births take place after the onset of the summer rains, from late December to early January. Calves can walk within minutes and run with the herd as little as five minutes after birth. Within a day they are capable of keeping up with the herd, and groups of calves can frequently be seen galloping and frolicking together as the herd moves about the veld.

The calving season is a time of plenty for the predators, however, with a large percentage of the newborn calves, lambs and foals falling victim to lions, leopards, cheetahs, hyaenas and jackals. As if to emphasize this point, we later see several jackals squabbling over the remains of a springbok lamb, and one runs off with the victim's head dangling from its jaws.

JANUARY 11. The dawn chorus of titbabblers, bleating warblers and redbilled buffalo weavers above our campsite has us on the move before it is fully light, but we are rewarded for the early start with our first sighting of an African wild cat. This nocturnal predator, not often seen in daylight, looks very much like a household tabby cat. And although not rare, it is regarded as seriously endangered because it interbreeds readily with domestic cats, thus destroying its wild bloodline. Usually shy and unco-operative – at least from a photographic point of view – the wild cat sits perfectly posed in the early morning light, watching us with curiosity.

Continuing, we see numerous springbok and wildebeest herds but the calving seems to have been suspended, possibly on account of the hot, sunny weather that has returned, and the only young we see are now several days old. The rains have brought a myriad new birds, an arrival which has brightened the landscape with unfamiliar colour. Although Etosha is a birder's paradise all year round, the summer months bring many migratory species such as European swallows, bee-eaters and rollers, white storks, avocets, lanner falcons, lesser kestrels and numerous other raptors, sandpipers and plovers, ruffs, and countless waterfowl.

JANUARY 12. It is still dark when we leave camp to meet our friend and sponsor, John Matterson, who has offered his aircraft for aerial photography, and when we take off it is barely light enough to make out the trees at the end of the runway.

Winging over the awakening countryside, we see numerous herds of gemsbok, wildebeest and springbok before we cross to the edge of Etosha Pan to look down on its empty whiteness. John sets a north-westerly course towards the Ekuma and Oshigambo deltas, the remnants of rivers that once flowed strongly into the great lake that was Etosha. Now shallow watercourses that flow from Ovamboland in the north, they empty their contents into the pan only in years of good rainfall. In very wet seasons, rains to the north and east of Etosha feed both of these watercourses and also the Omuramba Ovambo which flows into Fischer's Pan near Namutoni. Spilling into the pan at several points, the water spreads across the vast surface creating a sparkling wonderland which is attended by flamingoes, white pelicans, black-winged stilts and other waders; also by countless flights of ducks and geese, which gather to feed on nature's nutrient-rich plankton and algal 'soup'.

But now both the Ekuma and Oshigambo rivers are virtually dry, isolated pools the only remains of the heavy rains that fell a few days ago, and we are

The Oshigambo delta is a remnant of the river that once flowed into Etosha Pan.

unable to get the photographs we had hoped for. As the rising sun casts long fingers of light and shadow across the surface below, we turn south towards the distant twin hills at Helio, one of the few geographical landmarks in this featureless expanse.

As we leave the pan to return to the grassy fringes and sparse woodland surrounding it, we see two black rhinos galloping across the veld, startled by the sound of the aircraft. There are numerous other animals too: zebras, wildebeest, giraffes, gemsbok and ostriches, as well as the ubiquitous springbok.

The hours pass in wonder and fascination, and we finally turn back towards the airfield at Namutoni as the sun arcs higher, turning the landscape below into a mosaic of greens, blues, yellows and browns; the glaring white expanse of the pan to the north a harsh and striking contrast.

JANUARY 14. Back at Halali we leave camp after an early breakfast to encounter three hyaenas basking in the morning sun. They are sleek and in prime condition, jowls still bloodied and bellies bulging from a recent feast. Further on, we pass several large herds of zebras drinking at the Rietfontein waterhole, and then take the circular route that swings northwards, away from the main road. There are hundreds of springbok scattered across the plains, many of the ewes suckling young that cannot be more than hours old. Three newborn lambs, still damp and helpless, lie near the roadside. Hopefully, we scour the herds through binoculars for any females showing signs of imminent labour.

The females usually move a short distance away from the rest of the herd before giving birth, but we have no success in spotting any. Turning back, we drive slowly past the animals once more, and are about to drive on when a springbok feeding nearby turns her back on us, revealing two tiny hooves protruding from her vulva. She browses a while longer, then crosses the road, closely followed by a springbok ram which appears to be encouraging her to move further away from the herd. We follow the pair for some distance, intrigued by the interaction between the ram and ewe. The ram makes a concerted effort to keep the by now considerably agitated female away from other animals, and eventually they settle in an open area several hundred metres from the rest of the herd. The ewe continues browsing while the ram struts about, alert and attentive. From time to time he makes clumsy attempts to mount the ewe, who has still progressed no further into labour.

After about 90 minutes the springbok lies down and several spasms and contractions rack her body. Soon the hooves are followed by frail legs, and then the lamb's head appears, still encased in the foetal membrane. The ewe stretches around and tears at the sac with her teeth, enabling the lamb to take its first breath of air, but then stands again and walks a short distance away where she feeds on a small shrub. The casual air with which the springbok behaves is astonishing, the lamb's head lolling from side to side as the ewe walks about. Our concern for the survival of the lamb is relieved when it lifts its head and moves its legs. We wait another hour before the female lies down again and further contractions ensue, all the while the male standing at her side, watching intently around and obviously guarding the labouring female.

With several further heaves and spasms more of the lamb emerges, half-out of the ewe now, and then, to our amazement, the lamb lifts its head, looks around at its new environment, and nibbles at a bush alongside. It takes another quarter of an hour before the birth is completed, when the ewe stands up and eats the sac encasing the lamb's body. She then roughly licks it dry. The ram stands proudly by, watching the performance and then both move a short distance away to graze.

The lamb lies motionless, then makes several feeble attempts to stand, but it takes more than half an hour before it succeeds in gaining its feet for the first time, and then very unsteadily. The ewe watches closely, but appears reluctant to allow her lamb to suckle immediately, backing away every time it attempts to reach her teats, and brusquely setting about completing the grooming. About 15 minutes later she relents and stands still for the first feed, the lamb having some difficulty in finding the teats, nibbling at its mother's flanks, legs, and chest before voraciously succeeding.

The lamb eventually takes its first faltering footsteps, slowly gaining strength and confidence as it follows its mother a few metres away, but then lies down, hidden among a clump of grass and shrubs. A second ewe and her lamb move closer, and the two newborn springbok settle together. Springbok usually hide their young for the first day or two of life, and it is only after this period that the youngsters join the herd and are able to move with it, quickly strengthening and gaining in mobility. Although the young are supposedly protected from detection by predators by the lack of body scent in the first days of life, only a small percentage survive to maturity.

Later that afternoon we are stunned and saddened when we watch a cheetah chase and catch this very lamb, playing cat-and-mouse-like with it before paralysing it with a bite to the back of the neck. It then eats it alive, while the stricken ewe bleats pitifully nearby, craning to see what is happening.

JANUARY 18. After a hazy, wet morning of dull grey skies, we wait until late afternoon before heading towards the pan. Between Rietfontein and Salvadora a leopard is crouched in the grass, its attention fixed on a herd of springbok feeding nearby. With painstaking care in its stalk, it moves slowly towards the unsuspecting antelope, ignoring the vehicle when we drive closer. We then see a springbok lamb – the intended victim – lying 'hidden' in some brush about 25 metres from the leopard.

Still the cat inches slowly forward, hugging the ground, its tail extended straight out behind it, making use of every scrap of cover between it and the springbok. Suddenly, it bursts into motion and, in a blur of black and gold, streaks over the intervening ground. Too late the young springbok tries to escape, barely having had time to gain its feet when the charging leopard bowls

A springbok lamb lies hidden amid scrub and grass tufts.

violent momentum. For timeless seconds the leopard lies still, the springbok struggling and bleating in its grip, before it stands and kills it with a bite to the base of the skull. Although only a few brief seconds have gone by, the time seems to have passed in slow motion. The leopard drops its prey at its feet, and lies panting, looking in our direction with hot, yellow, satisfied eyes. It moves a short distance away, and does not feed until some time later.

JANUARY 22. After an uneventful early morning's drive in the Namutoni area we encounter another springbok giving birth on the open grassy plain near Klein Namutoni. This time, however, we see none of the drama of the previous lambing and the delivery takes less than 30 minutes. We watch and photograph the whole birth and witness the lamb taking its first, wobbling steps, before suckling from its mother.

Later, we pack our vehicle and travel to Beisebvlaktes, a large pan in the far north-eastern corner of the Park, where we plan to spend a few days.

The countryside is lush and green now, and new growth has rejuvenated the landscape wherever we look. Numerous wild flowers have appeared, and at Beisebvlaktes a carpet of yellow, daisy-like *Hirpicium gazanioides* lends an enchanting, park-like appearance to the land. Looking at the emerald-green plains scattered with flowers and dotted with wildebeest, springbok, zebras and numerous birds, it is hard to believe we are really in Etosha.

We see a lone black rhino bull which issues a sharp snort and crashes away through the brush, and then we find a group of banded mongooses at their burrow in an old termitarium, where we spend several hours. There are a number of babies, which eventually leave the burrow and sit warming themselves in the afternoon sun. Banded mongooses are more gregarious than many others in the family, and we estimate that there are about 40 of them living here. Excellent hunters, these small predators are legendary killers of large venomous snakes such as cobras and mambas, though their diet is primarily comprised of insects, grubs, eggs and small reptiles, as well as the fledglings of ground-nesting birds.

Later, as the sun casts a warm glow over the open flats, we collect a basketful of tasty beefsteak, or *Termitomyces,* mushrooms for dinner, and settle down to enjoy the peace and tranquillity of this isolated corner of the Park.

FEBRUARY 3. It rains almost every day now: dank, gloomy weather, dull and depressing. To escape the grey days, we have taken to working at night, seeking out the nocturnal creatures which are rarely seen by day, and are privileged to have had our eyes opened to a different side of Etosha entirely.

Driving down the Elandsdraai track we see numerous lesser bushbabies, tiny furry primates with a mass of less than 200 grams. Nocturnal and arboreal, they take flying leaps through the tree-tops, their eyes glowing brightly in the beams of our spotlights. Bushbabies feed on a combination of the gum exuded by certain tree species, and insects; and now they are foraging through the mopane trees in pursuit of moths and other insect prey.

Later we see a small-spotted genet, another of the small nocturnal cats, and follow it on foot as it hunts in the undergrowth. The cat eventually catches a mouse, and scurries out of sight to feed.

FEBRUARY 7. Returning from Okaukuejo camp, where we have been to collect mail and make telephone calls, we spend time photographing a Cape penduline tit nest at the roadside. These nests are unique, incorporating a false entrance which is designed to mislead predators attempting to raid the nest for

With a final heave, a springbok ewe thrusts her newborn free.

eggs or chicks. The cock bird returns on several occasions with small grubs and other insects, first poking its head into the false doorway in a display meant to mislead any observers, then unobtrusively pulling down the flap that covers the real entrance, where it passes the food to the occupants. The nests can hold as many as 10 chicks at a time, though this number would probably represent the offspring of more than one female.

Much later we are fortunate to photograph an African wild cat hunting. We also encounter a porcupine, but it flees as we stop the vehicle, disappearing into thick bush. Numerous genets are about, as well as an aardwolf and a brown hyaena, both uncommon in Etosha today.

FEBRUARY 10. Namutoni. We leave camp after dark and drive towards Aroe, an area where we hope to find nocturnal animals such as aardvark, porcupines or aardwolves. The plains are alive with springhares, which look like miniature kangaroos as they hop about the countryside. These large rodents – not related to hares at all – are strictly nocturnal, emerging from their burrows only well after dark, when they begin to feed. They are grazers, and live almost entirely on grass. We see literally hundreds of them on the plains that fringe Fischer's Pan, their activities quite a comical sight to watch.

Crossing the causeway that spans Fischer's Pan near Two Palms, we find a number of dinner-plate-sized African bullfrogs in the road. These frogs, the largest in southern Africa, may grow to a length of 200 millimetres during a lifespan of at least 28 years. They only emerge from their underground 'cocoons' after heavy rains fill the dry pans and vleis, and are able to remain underground for great periods of time if suitable rains do not fall. In fact, several years may pass in Etosha without the frogs being seen.

We watch bemused as two of the huge amphibians lumber across the road and disappear into rainwater pools alongside the track; and then continue on our way towards Aroe with the intention to return to the frogs at first light. Rounding a bend in the road we encounter a porcupine drinking from a puddle, and for the next 40 minutes we follow it on foot as it forages across the flats, apparently unconcerned by our attentions.

A Cape penduline tit peers warily from its nest before exiting.

FEBRUARY 11. Two bullfrog males are fighting for their territories – favoured pools of water – where they will mate with several females during the course of the day. The frogs are extremely aggressive, and are capable of inflicting serious wounds with two prominent tooth-like projections on the lower jaw. Several of the frogs display evidence of recent encounters in fresh wounds, and also the scars of old battles on their heads and shoulders. Once dominance has been asserted the loser generally moves away to seek another pool elsewhere. All the while a deep, booming sound, like distant off-road motorcycles, echoes across the pools as males court the awaiting females.

Several females, which are much smaller than the males, enter a pond and vie with each other for the male's attentions. Showing great stamina, the male satisfies one after another. Each female is capable of producing as many as 3 000 to 4 000 eggs which will take two days to hatch. Curiously, these bullfrogs are known to show parental care, males having been recorded digging canals to run water to pools in danger of drying up, and on occasion going so far as to build dams to retain water for the tadpoles. The tadpoles take 18 days to metamorphose, after which time the countryside comes alive with small green and yellow frogs.

Bullfrogs will eat anything that moves. They are also cannibalistic, fellow bullfrogs probably making up a major portion of their diet.

We spend most of the morning photographing the frogs, and then watch them burrowing backwards into the sodden earth. By midday not a single bullfrog is in sight, and we return to Namutoni.

FEBRUARY 14. Although we search for them night and day, we find no further signs of the bullfrogs, and now the rainwater ponds in which we saw them are almost dry. It is astonishing that the frogs emerged for a brief period of about 12 hours, during which they fed and mated before returning to their cocooned state underground; and we ponder the fate of the tadpoles which now throng the fast-drying puddles. It seems likely that unless further heavy rains fall soon, this season's progeny will not survive.

Later in the day we return to Halali, where we find the plains desolate apart from several large herds of springbok. The zebras and wildebeest have been trekking west for the past two weeks, and all that remains are a few old, territorial wildebeest bulls staking out prime ranges in anticipation of the returning herds later in the year. This may give these bulls a slight advantage over other males which will have to start from scratch in claiming territories. It appears that dominant wildebeest bulls in Etosha defend individual territories, to which they attract females prior to the rutting season. This process sees them rounding up a harem herd which the bulls then try to prevent from leaving until after the rut. They defend these territories from neighbouring bulls with threat displays and horn sparring, though physical injury seldom occurs.

These few old bulls look decidedly forlorn as they patrol their small ranges. We get to know several by sight, and usually find them resting through the heat of the day in the shade of the same prominent trees.

Many flowers are in full bloom, and Rietfontein is spectacular with a carpet of yellow 'duiweltjies' (*Tribulus zeyheri*), though there is scant game about to benefit from the nutritious grazing it offers.

FEBRUARY 15. Heavy rains during the past few nights have left sheets of water over huge areas of Etosha Pan, and French film-maker Alain Degré has offered us the use of his microlight aircraft to capture the sight on film. Although the day starts out grey and miserable, it begins to clear in the afternoon and we take off from Halali airfield, anticipating a colourful sunset from the air.

Towering storm clouds, broken by patches of clear sky, make a glorious spectacle, the whole bathed in a spectrum of colours, from golds, oranges and yellows to intense reds and pinks. Reflections sparkle and glitter in ethereal beauty. The slow-flying microlight makes the ideal photographic platform, and we swoop and soar like eagles in the wind. The magic changes with each passing minute, and we stay over the pan until the last possible moment, then race the rapidly approaching darkness back to the airstrip.

FEBRUARY 17. Most of the large herbivores are concentrated in the Okaukuejo area now, and we decide to spend a few days there – a final visit, for our sojourn is drawing to a close. Soon we will have to pack to leave Etosha. Like all good things it will end.

Huge herds throng the plains west of the restcamp, and appear to be moving westwards still further. Zebras, wildebeest and springbok teem in the thousands, and two different lion prides attend the masses, not far apart.

At Sprokieswoud the moringa trees are topped with crowns of feathery leaves, and bright green growth has renewed the fire-ravaged grasslands. Reptiles of all kinds are about, countless chameleons, monitor lizards, terrapins and tortoises, and a fair number of snakes slither across the roads as we pass. There is a new vigour in the land and all its creatures, and we are happy to be a part of it.

FEBRUARY 23. We return to Halali, stopping at Rietfontein, more out of habit than in expectation, for there is water everywhere and none of the animals needs come to this spring. To our amazement a breeding herd of 32 elephants emerges from the trees and approaches the water. They drink deeply, then splash and roll with abandon. The animals' presence here is unusual, for we have not seen elephant in this part of the Park in almost two months.

After almost an hour the herd files away, trumpeting a last farewell that will stay with us as we prepare to leave our home – for that is what Etosha has become during the enchanted months we have stayed here.

RESEARCH & MANAGEMENT

Few visitors to game reserves and national parks ever come into contact with, or understand the work undertaken by research and management officers. But it is their work and devotion to the cause of conservation that enables visitors to observe and, hopefully, appreciate wildlife in its natural state.

Wildlife management today entails more than erecting a fence around a proclaimed game reserve, keeping the animals in and poachers out, though this is still a central function of the modern game ranger. Park management is a precise science, with rangers mostly in possession of a minimum qualification of a three-year diploma in wildlife management or a bachelor of sciences degree, for management is heavily research orientated.

According to the Nature Conservation Ordinance (No 4 of 1975), Section 13, Etosha National Park '... shall be a game park for the propagation, protection, study and preservation therein of wild animal life, wild plant life and objects of geological, ethnological, archaeological, historical and other scientific interest...', and further, as defined in the Masterplan for Etosha National Park, should have as its objectives 'the maintenance, and in special cases the increase, of biotic diversity of the local biota, and the sustained utilization of its resources for the benefit of humanity'.

The task of carrying out these objectives is the responsibility of the Park's management, which is divided into three divisions: tourism, research, and management. The tourism division operates semi-independently of the others, maintaining and operating the restcamps and other visitor facilities, but has no say in actual wildlife management. Decisions regarding the management of the Park's plant and animal life, including all aspects of its ecology, are made on a collective basis by the research and management divisions, in compliance with the Masterplan drawn up for the Park.

Although the romantic image of a game ranger patrolling the veld on horseback is still reality in Etosha, this is but a small part of a ranger's routine. Allan Cilliers, Chief Conservation Officer (Management), maintains that the game ranger of today is in effect a highly qualified 'ecological manager', for his duties include such diverse responsibilities as the monitoring of all flora, fauna and water resources, erosion control, boundary patrols and fence maintenance, law enforcement and even extension work such as rural development and liaison with neighbouring farmers, who may often have problems with animals crossing Park borders, damaging crops or killing livestock. Add to this the mundane day-to-day maintenance of water installations such as windmills and diesel pumps, road and firebreak construction and maintenance, and a task few enjoy, the control and culling of problem animals, and the modern-day game ranger has a busy time.

So the often gruelling tasks such as assisting with game capture operations, or week-long horseback patrols through remote country come as a relief from other more mundane routines, and give the rangers a brief respite in which to get back to the basics of their profession.

THE ETOSHA ECOLOGICAL RESEARCH INSTITUTE

The Etosha Ecological Research Institute (EERI) at Okaukuejo is one of only four special research institutes established within national parks in Africa. The other three are located at Gobabeb in the Namib-Naukluft Park (also in Namibia), Skukuza in the Kruger National Park in South Africa, and Seronera in Tanzania's Serengeti National Park.

Head of the EERI, Dr Malan Lindeque, says the unit is the backbone of

Horse patrols are still an important facet of the rangers' routine.

Radio collars help to keep track of animals under research.

research into specific projects both by staff members and outside research biologists, and functions specifically to provide scientific data to ensure the proper management of the Park: 'We provide a monitoring service to enable Etosha to be managed according to modern, sophisticated techniques, and try to concentrate on predictive research, so that we can actually prevent things from happening rather than treat them after the event. ... We are trying to get away from problem orientated research, and to this end we also encourage outside researchers to work with the Institute.'

Lindeque maintains that a priority in Etosha is monitoring the vegetation, for this is the basis of all other life in the Park. 'The vegetation can change very quickly here, and constant monitoring is essential in a semi-arid environment such as we have. Often changes are irreversible, so I consider this one of our most vital functions. And of course, monitoring climatic and soil conditions is an important part of this.'

The Institute also keeps a check on all the important large game species, for in today's game reserves with their game-proof fencing and other artificial restrictions on animal movements, wildlife populations and numbers require scientific control. Before the fencing of Etosha, its wildebeest and zebra populations each numbered about 30 000. Today they are down to 3 000 and 4 500 respectively, primarily due to the closing of migration routes. When these populations declined so fast it was initially thought to be the result of too large a lion population in the Park, and research was done into possible methods of population control.

In a widely publicized research project undertaken by former chief biologist Dr Hu Berry in the early 1980s, lionesses in Etosha received hormone-impregnated implants, or slow-release capsules which prevented them from conceiving for a minimum of three years. Considerable work was undertaken at the same time into the social implications of these implants, for lions, unlike other cats, have complex and close social ties. Although the project was considered a total success, it has never been required to be fully implemented, although the technique could be used if ever considered necessary.

HORNS AND IVORY

Two of the major species requiring constant monitoring are the black rhino and the elephant, both targets of commercial poaching gangs wanting horns and ivory. The elephant population averages about 1 750, and has shown a consistent decline since 1983. Numbers vary, however, between the wet and dry seasons, the animals dispersing when water and food are plentiful. Many migrate northwards only to return in the dry months, attracted by a steady supply of water in the Park. Such nomadic habits put the animals at considerable risk, as once outside the Park they are unprotected and at the mercy of hunters and farmers. It is interesting to note that elephants were totally absent from Etosha for 70 years, and only began to re-colonize the area in the late 1950s, taking advantage of the protection which it offered from the areas beyond its borders.

The black rhino is one of Africa's most seriously endangered wild animals, the population having been decimated by illegal hunting over the past two decades. Twenty years ago it was estimated that 65 000 black rhinos roamed the continent; today that figure stands at a shameful 3 000-3 500, and is still decreasing. Namibia, and Etosha National Park in particular, stands as a bright beacon in the preservation of this species, for here, unlike most other places

Etosha rangers have devised a unique method of monitoring black rhinos at night.

which the animal still inhabits, black rhino numbers are increasing steadily. From a 1970 estimate of 48 rhinos within the Park's borders to a 1990 figure of about 340 animals – 10 per cent of the world population – Etosha has made great strides towards the conservation of this species.

Namibia's black rhinos are of the sub-species *Diceros bicornis bicornis*, the famed desert rhino found also to the west of the Park in the rugged Damaraland and Kaokoland regions, and, it is believed, the same sub-species that once occurred as far south as Cape Town's Table Mountain. Namibian conservation authorities are today working closely with their South African counterparts to reintroduce the sub-species to game reserves in the northern parts of the Cape Province.

Such reintroductions and relocations form an important part of rhino conservation strategy. The ongoing capture and relocation of black rhinos within Etosha from isolated, high-risk areas to places deeper within the Park where poaching is less of a risk, combined with regular monitoring and identification

Elephants have only returned to Etosha over the past 30-odd years.

programmes, have proved successful in maintaining the population. For example, in 1989 as many as 23 black rhinos were poached within the Park, but by late 1990 the figure had been reduced to two.

The technique used to monitor rhinos in Etosha is unique to the Park and worth explaining. Allan Cilliers describes the procedure in a research document on the subject: 'Water-holes were covered systematically during the dry months which are usually from May to October when no open veld water occurs. Teams of observers are placed at water-holes for three consecutive nights, during full-moon periods, starting six days before the date of the full moon.' The intention is to age and sex the animal before it reaches the water and starts drinking, then to photograph it from the front and side using carefully placed piles of rocks as distance markers.

Using such techniques, Etosha's rangers have compiled identification dossiers on some 300 of the Park's estimated 340 black rhinos.

Radio telemetry, long an important aspect in the monitoring of other game species, has now been introduced into the battle to save rhino, and we joined a rhino capture and relocation operation to see how this was done. Several teams posted overnight at waterholes in the designated capture area are instructed to follow the spoor and track down the rhino at first light.

When we awoke shortly after dawn on the first morning, two teams were already on the trail of rhino. Some hours later, one team had caught up with one of the animals, and the helicopter was despatched carrying the veterinary officer who was to dart it. We made our way to the area in a 'chase vehicle' and arrived in time to see the helicopter swoop down, hover scant metres above the thorn scrub vegetation, and then release the dart into the animal from the air. After a frantic, crashing headlong run through the bush lasting several minutes, the rhino went down.

Veterinarian Dr Pete Morkel rushed to the inert animal to monitor vital signs and ensure that it was lying comfortably. After discerning that the rhino was suitable for relocation – to a safer area within the Park – the task of inserting a radio transmitter into its horn began. This device, a compact cylindrical object about the size of a human thumb, was inserted into a hole drilled near the base of the horn. The hole was then neatly sealed with dental acrylic and a plug cut from the tip of the rhino's horn, making it impossible to see that the horn has been tampered with. Interestingly, conservation authorities believe it would be possible to track down a horn containing one of these transmitters, right to a poacher's cache.

While the rhino was drugged its ears were notched, to aid visual identification of the animal in future. It was then partially antidoted and loaded into the crate in which it was to ride to the release area.

POACHING

Apart from the successes achieved through relocation and monitoring, the establishment of a specially trained Anti-Poaching Unit (APU) in December 1988 has had a notable effect in reducing poaching incursions. The unit, well-trained, well-equipped and highly mobile, operates throughout the Park and throughout the year. Members often spend weeks at a time in the field, focusing their attention primarily on commercial poachers.

Says Allan Cilliers, who was instrumental in establishing the anti-poaching unit: 'We are not that concerned with subsistence poachers, the person who climbs over the fence to shoot a springbok or warthog in order to feed his

The monitoring of vegetation is an ongoing function of park biologists.

family. The guys we're after are the commercial gangs, often armed with automatic rifles, who come into Etosha and gun down 10, 20 animals at a time. Or who go after our rhino and elephant.

'A lot of the APU's success today depends upon undercover work. They rely a lot on the informer system, and fortunately we're getting a lot of information from the local population. At last it appears that many of the tribal headmen are realizing that wildlife is part of Namibia's heritage, and that they all stand to gain from preserving it.'

However, with rhino horn selling for as much as $20 000 a kilogram in the Far East, it is highly unlikely that poaching will abate, particularly as rhino populations further north in Africa are depleted or, in fact, poached out. As guardians of the largest single population of black rhino within a national park in Africa, Etosha's trustees have a weighty responsibility to the species, and indeed to mankind.

Desert elephants in Kaokoland have frequently fallen prey to poachers.

DISEASE CONTROL

Another major function of the Etosha Ecological Research Institute is the monitoring and control of disease, anthrax and rabies being the two major ones in the territory.

Anthrax is an infectious bacterial disease that affects warm-blooded animals, including man, causing fatal septicaemia. Widespread throughout Africa, and indeed occurring throughout the world, the disease has become prevalent in Etosha over the past 30 years, at one time being responsible for more than 50 per cent of the total recorded mortalities in the Park.

Anthrax has been diagnosed in game as diverse as elephant, giraffe, eland, wildebeest, plains zebra, gemsbok, kudu, springbok and ostrich, mostly herbivorous animals although instances in which cheetah have been affected are on record. It would appear that carnivores such as lion, leopard, hyaena and jackal are immune to the disease, for although often recorded feeding on anthrax-infested carcasses they have never in any known instance contracted the disease themselves. It has in fact been determined that most wild animals have a natural resistance to anthrax, which can be lowered, however, by factors such as stress, poor nutrition or vitamin deficiencies. The disease is highly contagious, and is contracted by inhaling or ingesting anthrax spores.

Veterinarians maintain that the erection of fences around Etosha has played a major role in increasing the incidence of the disease. Migratory species which under normal circumstances would have left the Park and so given the land a necessary respite during winter, are now prevented from doing so. The resulting year-round concentrations of game have led to over-grazing followed by weed and bush encroachment, and so also to the increased risk of disease. Another major contributing factor to the spread of anthrax is believed to have been the gravelling of Etosha's tourist roads, an undertaking which necessitated the digging of quarries, or gravel pits. In times of good rain these pits fill with water and so serve as water reservoirs, leading to a continued grazing presence and subsequent over-utilization of the land. The provision of water through artificial means has exacerbated the problem by discouraging the animals from migrating and so subjecting the land within a limited area to enormous pressure.

Lions and other carnivores have an immunity to the killer disease.

Scientists have recorded a definite relationship between the over-concentration of plains animals, causing over-grazing and vegetation degradation, and the spread of anthrax, and in an effort to control this a programme of disinfecting waterholes and filling-in disused gravel pits has been introduced.

Because of the threat posed by this disease, almost every natural death discovered in the Etosha National Park is tested for anthrax.

Dr Lindeque outlines briefly: 'Very little is actually known about the disease, although here in Etosha we have one of the very few specialized anthrax laboratories in Africa. ... Apart from ongoing monitoring, and disease-control tactics such as ensuring that the Park vegetation is not degraded by over-concentration of large herbivores, there's little we can do. It's doubtful if the disease will ever be totally eradicated from Etosha, or other game areas where it's present.'

Rabies too has caused many deaths in certain game populations, including lion, jackal and even black rhino. Kudu are highly susceptible to the disease, and a major outbreak in Namibia between 1979 and 1982 killed more than 100 000 of these antelope. Again the constant monitoring of the Park's vegetation is the priority with regard to these problems.

The responsibility and functioning of the EERI extends well beyond the borders of Etosha and its concerns are greater than the issues arising within its own area. The Institute is available to the public sector and to commercial farmers to assist and advise in conservation-minded soil usage. The Park, being what it is, offers an opportunity to compare what has happened to the ecosystems of north-western Namibia with different land usage. According to Lindeque, it can be proved 'that game can do far better and produce more meat per hectare than cattle, and do less damage to the land itself'. Lindeque maintains that farmers on Etosha's boundaries, rather than consider the Park a nuisance, 'should realize the value of having [it] in their middle. There is tremendous potential for the development of tourist facilities, private camps and parks, such as those around the Kruger National Park in South Africa.'

Both Lindeque and Cilliers agree, however, that the old Africa is a thing of the past: 'Now we have to actively conserve what is left.'

Only the scavengers benefit from the ravages of anthrax.

THE BIG CATS

Africa's legendary 'Big Five' are the five most sought-after trophies of the big game hunter or safari-goer. Four of these rate highly on Etosha's checklist, while the cheetah makes a more than adequate substitute for the one that does not occur in these parts, the buffalo.

Etosha's 'Big Five' therefore comprises the three 'big cats', lion, leopard, and cheetah, and the elephant and black rhino, all of which are readily observed during a visit of reasonable duration. The chance of seeing a particular species is better in certain areas than in others, though, which is a good reason to spread a visit over each of the three restcamps. Of all the members which comprise this group, it is the big cats that generate the most excitement.

LION

Although now restricted in their range to large parts of Africa and the Gir Forest in north-western India, lions have been traced back about 15 000 years through cave art and archaeological finds to parts of Europe, and Greek philosopher Aristotle wrote of lions in Greece as recently as 300 BC. The Crusaders recorded encounters with lions in the Middle East, and in fact the cats could still be found in much of the Middle East at the beginning of this century. Etosha, however, is still prime lion country, its wide-open plains, dense concentrations of prey and limited number of watering points, all strong incentives to their remaining here.

Lions are Etosha's major predators and number some 350-400. The only social cats, they are usually encountered in prides of five to 15, though prides of more than 40 have been recorded here in the past. Because lions are predominantly nocturnal, most daylight sightings of these magnificent creatures find them at rest, either asleep in the shade of a tree or shrub, or sprawled sunning themselves in the early mornings. It has been estimated that lions spend about 20 out of every 24 hours resting, and recent research has shown that the majority of hunts and kills take place at night.

Male lions are considerably larger than the females, with a mass ranging between 150 and 240 kilograms, compared with the female's 120-180 kilograms. With the male's chief role in a pride being territorial defence and defending kills against hyaenas and other opportunists, his added size is a distinct advantage. This may also explain the existence of the lion's mane, for it gives the appearance of even greater size without the disadvantage of added bulk.

In spite of being smaller, the females do most of the hunting, though after making the kill generally withdraw or are driven off, so that the males can eat first. The younger cubs sometimes join in the feast, their 'young smell' probably protecting them from attack by the males (McBride, 1990).

When hunting, lions usually rely on stealth, and will either lie in ambush or use available cover when stalking their prey. Although they prefer to attack from close range, they have been recorded in high-speed chases over as much as a kilometre, and are capable of hauling in even the fastest of antelope species, such as tsessebe and hartebeest (McBride, 1990).

Lions are opportunistic feeders and frequently drive other predators off their kills; they will readily eat carrion even when this is days old. In a study conducted in Tanzania's Ngorongoro Crater area it was recorded that more than 80 per cent of the lion's diet comprised carcasses of animals killed by hyaenas. Equally, however, groups of hyaena frequently chase lions from their own kills, particularly when there are no adult males present to defend it, gainsaying the hyaena's traditional reputation of skulking cowardice.

In Etosha, zebras and wildebeest appear to be the preferred diet of lions, although seasonal congregations and migrations play an important role in prey selection and availability. The big cats will actually feed on anything available, from reptiles and rodents, birds and hares, to giraffes, rhinos and even elephants. Fortunately, Etosha's predators, including leopards, hyaenas and jackals, appear to have an inherent immunity to the killer disease anthrax, and feed on the remains of anthrax-infested carcasses with impunity.

While most hunts do take place at night when cooler conditions combine with the cover of darkness, lions will occasionally kill opportunistically even in the hottest part of the day; they may also be seen hunting in daylight hours in the cold winter months and on overcast or rainy days, and it has recently been shown that they enjoy far greater success in cold or cool conditions.

Lion young are born throughout the year after a relatively short gestation period of between 100 and 115 days. Studies show, however, that females within one pride may come into oestrus at a similar time, which results in synchronous births. A litter usually ranges from two to six cubs, and the cubs will suckle indiscriminately from any lactating female, a behaviour that in the event of a lioness dying may ensure that her cubs survive to strengthen the pride. At birth cubs are very small and undeveloped, and there is typically a high mortality rate. Research has indicated, however, that mortality may be connected to lion population densities, a higher survival rate being recorded when territories are sparsely populated.

In Etosha, lion densities appear to be highest in the Halali and Okaukuejo areas, with waterholes such as Goas, Rietfontein, Charitsaub, Gemsbokvlakte, Ombika, Okondeka and Okaukuejo all noted lion-watching sites, depending on the season. The best time of the year for lion sightings is naturally the dry season when vegetation cover is sparse, and when watering points are in high demand and form prime hunting bases for the predators. It is worth bearing in mind that a pride of lions seen sleeping quietly in the shade of a tree at midday is likely to remain there all day, and come late afternoon will begin to awaken, playing and grooming each other, or even initiating a hunt.

CHEETAH

Haughty, imperious, proud, aloof and elegant are all words which aptly describe the cheetah, the fastest of all the land animals, and capable of speeds up to 110 kilometres per hour over short distances.

Easily distinguished from most of the other cats by its distinctive markings

and lithe, rangy build, its small head, and short, rounded ears, the cheetah is differentiated from the similarly spotted leopard by the bold black stripes that run from the eyes to the corners of the mouth. These stripes combine with the cheetah's black lips and dark mouth to emphasize facial expressions which are important for communication. Cheetahs have long, heavy tails which they use to good effect when chasing prey, throwing the tail as a counter-balance when cornering sharply. Relatively scarce in Etosha nowadays because of an unexplained slump in their numbers in recent years, these cats are still seen fairly frequently because of their apparent preference for more open country, although they do readily inhabit mopane and mixed savanna woodlands.

While generally considered to be solitary animals, the males frequently form associations with other males, often litter siblings, and females are most commonly seen in the company of their offspring, which remain with her until maturity. Unlike lions, however, they are essentially solitary hunters, relying on sight, stealth and speed to stalk to within about 30 metres of their prey before running it down with a short, fast chase. Although their physiology has evolved for this specialized form of hunting, a slender build and supple spine allowing them to take exceedingly long and fast strides, they lack stamina and can maintain full speed over short distances only.

Cheetahs rarely attack any animal that stands its ground, relying rather upon a high-speed charge to stampede the intended victim, for it is easier for them to overbalance and knock down a running animal than a stationary one. Once they have achieved this, usually by pulling a leg from under the moving animal, they kill their prey by suffocating it with a vice-like grip on the throat or over the mouth and nose. This hunting technique, and their preference for more open country, means that cheetahs fill an ecological niche different from those of the other major predators. Cheetahs generally hunt late in the morning and early in the afternoon, when predators such as lions, hyaenas and leopards are usually asleep, but even so will drag their kills under trees and shrubs in order to avoid the unwelcome attentions of vultures which themselves may chase cheetahs from a meal, or possibly arouse the interest of other predators. Springbok constitute the major prey of cheetahs in Etosha, though impala, ostriches, springhares and scrub hares, along with numerous smaller rodents, are frequently preyed upon. They will also attack the young of bigger herbivores, such as zebras, wildebeest and the larger antelope.

Unlike the other cats, cheetahs cannot fully withdraw their 'toenails' or claws, these being more dog-like in appearance. Males are slightly heavier than females, averaging between 40 and 60 kilograms compared to the female's 35-50 kilograms. Like the leopards and lions, cheetahs have a gestation period of about three months, and bear litters of two to five, though sometimes as many as eight cubs are born. Blind and helpless at birth, the cubs have a dark blue-grey mantle of hair on their backs, which they retain for three months. This aids in camouflaging them from rival predators, and from a distance gives them the appearance of a honey badger, renowned for its bravery and ferocity. There is a high mortality rate among cheetah cubs, but if they do survive to adulthood, the life expectancy is about 15 years.

Although Etosha's cheetah population fluctuates considerably, present estimates range between 50 and 100 animals. However, elsewhere in Namibia these predators are numerous, and are regularly shot as vermin by stock farmers, among whom they have a reputation as wanton killers of livestock. Cheetahs usually range over large areas, and according to research carried out in East Africa, have a home range of about 800 square kilometres. The best place to look for these cats in Etosha is generally on the open plains around

Leeubron-Adamax near Okaukuejo; also on the Halali plains, in the Springbokfontein vicinity, and around Namutoni, although patterns change from year to year according to game movements. It is worth checking with the tourist office at the nearest restcamp for the current trend.

LEOPARD

Solitary, secretive and elusive, the leopard represents the high point of a trip for many visitors to Etosha, possessing a mystique and aura impossible to describe. The largest of Africa's spotted cats, the leopard is sturdier and more solid than the cheetah and its black spots are clustered in rosettes which cover the body, no two leopards having identical markings. The background colour varies from a rich golden-yellow to off-white.

A highly adaptable and secretive nature has ensured the leopard's survival over a large part of its original range, and even today it is frequently reported from semi-urban and built-up areas. Being so elusive, however, censussing is difficult and leopard numbers in Etosha are unknown, though conservative estimates based on sightings by tourists and park rangers put the population at more than 100.

Leopards hunt using a combination of stealth, speed and strength, and primarily at night when they can approach to within close range of their prey before making a final, fast rush. They rely more on pouncing and subduing their victim than on chasing it, however. Death is by means of a powerful bite to the base of the skull or nape of the neck, but they will occasionally suffocate their victims in much the same manner as the cheetah. Immensely powerful, leopards frequently drag their prey into trees, in some cases even animals far heavier than themselves; we were told of a leopard that carried a newborn giraffe calf into a tree, a load of at least 90 kilograms. In Etosha leopards prey predominantly on the smaller antelope species such as dik-dik, steenbok, springbok and impala, but also hunt warthogs, birds and rodents, and even insects. They will also feed on carrion and have frequently been observed feeding at anthrax-ridden elephant carcasses.

Males are generally heavier than females, and weigh between 50 and 90 kilograms while females have a mass of between 30 and 60 kilograms. Litters range from one to six cubs but the average is two or three, and the cats have a life expectancy of 20 years.

Although solitary animals, they may occasionally be seen in groups of two to four, usually comprising a female and her immature offspring, although on very rare occasions an adult pair and its young may be encountered. The young, weaned after about three months, start accompanying their mother on hunts from the age of about four months and remain with her for up to two years. Affectionate reunions between a female and her adult offspring may take place after they have parted company.

Leopards are primarily terrestrial animals, particularly so in Etosha, though they are good climbers and may occasionally be seen lying at rest on a convenient tree branch, or using it as a lookout. However, probably as a consequence of Etosha's primary vegetation types and a dearth of suitable trees, most leopards are seen on the ground.

In Etosha, the Namutoni area has the best reputation for leopard sightings, particularly Klein Namutoni, Kalkheuwel and the two Okevi waterholes. We also had considerable luck with leopard in the Halali area, particularly at the Rietfontein and Goas waterholes.

ETOSHA
A WILDLIFE PORTFOLIO

Etosha is renowned for its elephants. Even though the population is relatively small, numbering some 1 750, the animals are readily seen in the dry season, gathered in large groups at the Park's waterpoints where herds of up to 100 queue to drink and bathe (*left*). When the summer rains begin to fall, however, the herds disperse, migrating to areas far to the north of Etosha, and visitors to the Park are fortunate to encounter even one elephant during their stay. The flap-necked chameleon (*above*) is the largest and most common of those found in Etosha. It may often be encountered making a spasmodic journey across a road, its jerking progress accompanied by constant vigilance as its eyes, which can move independently of one another, swivel to keep a close lookout for enemies such as birds of prey.
(*Previous spread*) Gemsbok in full flight across Etosha Pan.

A young elephant rolls in the dust, coating his skin after a cooling shower at the Mushara waterhole in the north-east of the Park. Elephants regularly bathe in both water and dust in an effort to cool their huge bodies and keep them free of parasites such as ticks and biting flies. While the adults generally go about these ablutions with a certain dignity, the young do so with youthful abandon.

Pure ecstasy. A warthog sow – males have two pairs of wart-like protuberances on the face – scratches her rump on a convenient rock after a thirst-quenching drink at Olifantsbad near Okaukuejo restcamp. Although not abundant in Etosha, these uncomely creatures are widely distributed throughout the Park and can usually be seen at watering points during the heat of the day.

(*Previous spread*) A group of giraffes etched in sharp silhouette beneath sunset-tinted storm clouds in late spring. Giraffes are both intensely curious (*left*) and constantly alert, attributes that combine with a nervous disposition to give the appearance of vulnerability. In actual fact, these animals can be determinedly aggressive and there are numerous accounts of them having turned on attacking lions, inflicting serious damage and even death by means of flailing kicks with their powerful forelegs.

The sun appears to be setting on Etosha's blue wildebeest population (*above*); where once more than 30 000 of these animals roamed, now fewer than 3 000 remain, one of the horrible consequences of fencing Etosha's boundaries. This undertaking meant that important migration routes were severed, cutting the animals off from their traditional summer grazing grounds. The strikingly elegant avocet (*right*) is one of many waders that flocks to Etosha in the wet summer months.

A cheetah interrupts its meal (*left*), its penetrating glare indicative of its displeasure at being disturbed. Although limited in number, cheetahs are among the most readily seen predators in Etosha because of their preference for hunting the open plains during daylight hours. Ranking fairly low on the hierarchical scale, they frequently lose their kills to other predators such as lions, hyaenas, leopards and even jackals and vultures. Cheetahs are generally solitary animals, although small family groups comprising a mother and her immature offspring are often encountered. Springbok (*above*) are the Park's most numerous species and are the dominant prey of cheetahs here, both animals preferring the wide-open plains and grasslands.

A trio of juvenile cheetahs (*above*) gives chase as a scrub hare bounds across the fringes of Etosha Pan, jinking in an attempt to throw off its pursuers. Cheetahs can reach a speed of 115 kilometres an hour over short stretches, and use their long, heavy tails as a counterbalance when turning sharply. In this instance, the hare's speed and agility gave it the edge, and a crevice among some rocks provided a convenient hide-out. Primarily nocturnal, scrub hares (*right*) lie up in 'forms' during the day, usually under low scrubby bushes, emerging at sundown to feed. When lying up the hares keep their distinctive ears folded flat on their backs and their heads pulled into the body, while the drab, cryptic colouring of the body blends with the colour of their surroundings.

It is only from the air that one can truly appreciate the size of Etosha Pan, which covers more than 6 100 square kilometres of the 22 275-square-kilometre Park. Far from being a trackless wasteland, the pan is criss-crossed by game paths such as this used by a family of zebras (*left*). The paths are worn by animals crossing from one side of the pan to the other in search of greener pastures. The lesser galago, or bushbaby (*below*), is so named because of its human-like cry which may frequently be heard in the bushveld at night. It is strictly a nocturnal animal, emerging after dark from its nest – usually located in a hollow tree trunk – to feed on insects and the gum exuded by some trees. The striking African spoonbill (*right*) feeds in shallow water where it probes the soft mud with its spatulate bill, or sweeps it from side to side in search of water insects.

ETOSHA: A WILDLIFE PORTFOLIO

Backlit by the early sun of dawn, an African wild cat (*left*) pauses on the return from its nocturnal wanderings. This species is relatively plentiful but is regarded as endangered because of its tendency to interbreed with domestic cats, so destroying its genetic purity. Like its domestic relatives, it feeds on small birds, rats, mice and insects. A magnificent full-maned lion (*above*) moves out at dusk to begin the night's hunt. Research has shown that lions hunt predominantly after dark, when cooler temperatures and the cover of night improves their success rate.

An early summer rainstorm sweeps across the Halali plains (*above*), bringing temporary relief to the parched lands after a long, dry winter. As in most of drought-prone Africa, the onset of the rains in Etosha is eagerly anticipated, and any delay in their coming can spell disaster. The small-spotted genet (*right*) is one of the small nocturnal cats that proliferates in the wooded areas of Etosha. These animals are excellent climbers and will readily take to trees in pursuit of their prey which includes small rodents, birds, reptiles and insects. They are regarded as terrestrial animals, however, as most of their activities take place on the ground. During the day they sleep in holes in the ground, in hollow logs or under dense bushes.

ETOSHA: A WILDLIFE PORTFOLIO

67

Africa's largest rodent, the primarily nocturnal porcupine (*above*) ranges over a large area on its nightly travels, seeking food such as roots, bulbs, fruit, berries and tree bark. Its predilection for the bark of *Moringa ovalifolia*, which predominates in Sprokieswoud, or the Enchanted Forest, has caused these trees serious damage. Crowned plovers (*right*) are common roadside dwellers and frequently lay their cryptically coloured eggs on road verges. Should a prowling predator approach, the mother bird will attempt to distract it by running away, calling loudly and shamming a damaged or broken wing, in the hope of distracting the predator from the nest bearing its eggs or chicks.

A dense bank of thunderclouds masses in the eastern skies near Namutoni (*left*), a promising sign of heavy rain later in the day. This eastern sector of the Park usually records a higher rainfall than other parts. An annual average of 550 millimetres is reflected in a greater diversity and density of vegetation types here. A notorious scavenger, frequently seen alongside vultures at carcasses, the repulsive and ungainly marabou stork (*below*) is also a predator of considerable note. It will readily stalk across the veld to feed on insects, frogs, lizards, snakes and rats. Its bare-headedness is an aid to cleanliness, for these birds frequently poke their heads into the carcasses upon which they are feeding. They perform a vital function in keeping the veld clean, and their powerful bills are often indispensable in opening up carcasses which vultures may otherwise have difficulty penetrating.

A red-necked falcon (*top left*) scours the countryside from its perch, seeking prey which includes rats and mice, lizards, grasshoppers, locusts and small birds. Etosha is home to numerous raptors, and is also visited by many other migrant birds of prey. A pair of gemsbok (*left*) is reflected in the still waters of a remote waterhole. These elegant antelope are true desert dwellers and are able to live without drinking at all, deriving moisture from their food, which may include certain wild melons. *Diceros bicornis* or *D. tricornis?* This three-horned black rhino (*above*), one of two such oddities which we encountered in Etosha, is undoubtedly a rarity though park rangers have recorded two others elsewhere in the reserve. Black rhino populations have been decimated by poachers who sell their horns for profit. The horns are used in the making of ceremonial dagger handles in the Middle East and in traditional medicines in the Orient.

Male lions, such as this pair which inhabits the Agab region of the Park, frequently form close bonds with each other. Such unions may develop between brothers from the same litter or merely between outcast nomads that link up to improve their success rate when hunting. Considerably heavier and hence slower than lionesses, male lions prefer to leave hunting to the females, but when solitary, have to rely on their own abilities.

A wildebeest calf, only minutes old, gains strength as it suckles from its mother while the rest of the herd waits patiently nearby. A calf is able to stand within minutes of birth and can run alongside its mother after about five minutes; after a day it is quite capable of keeping up with the herd, even at an extended gallop. Wildebeest have a gestation period of about 250 days and the calf weighs about 22 kilograms at birth.

Large communal spider webs (*above*) strung between small thornbushes and shrubs are seen throughout the Park, and can be home to hundreds of small social spiders of the genus *Stegodyphus*. This crimson dragonfly (*left*) is just one of countless insect species that brightens the Etosha landscape, though the dragonflies are among the most beautiful and conspicuous. They are most commonly seen flying rapidly up and down the edges of waterholes and ponds, where they both hunt and lay their eggs. Ground squirrels (*right*) show human-like affection as they greet one another in the early morning sun. They are highly social animals, and family groups may be encountered throughout Etosha along the roads, where they seem to have a preference for digging their warrens.

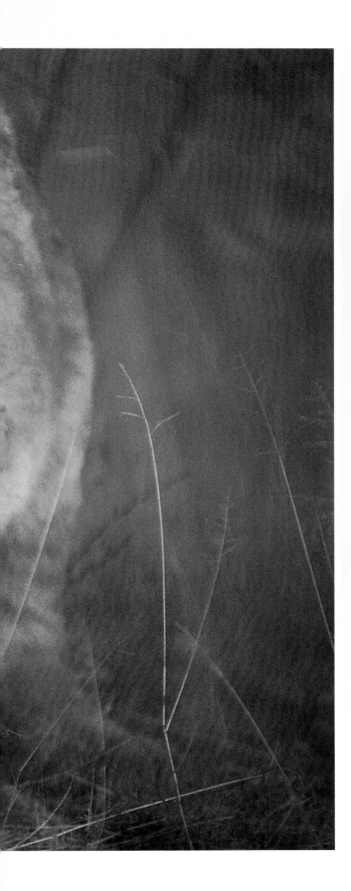

Two cubs from the 'Rietfontein pride' gnaw hungrily at the spoils of a springbok kill (*left*), having managed to snatch the remains of a leg from the squabbling adults nearby. Generally lion cubs have to take their chances at a kill, scavenging what they can from beneath the jaws of larger pride members. They have a high mortality rate, many dying of starvation when the pride is unable to kill often enough to satisfy the requirements of all its members. Lion cubs remain with their mothers for up to two years, at which age they begin hunting for themselves. A pair of red hartebeest engages in ritual sparring (*above*), an activity that can turn to vicious fighting between rival bulls during the rutting season.

Hartmann's mountain zebras (*above*) are found only in the more arid western parts of the Park, where they inhabit the rocky dolomite kopjes. They are easily distinguished from the more common Burchell's zebra by the clear striping all the way down their legs, and by their plain white bellies. Cape foxes (*left*) are among the rarer of Etosha's game species. Being predominantly nocturnal, they are rarely seen by day although they may be encountered warming themselves outside their burrows in the early morning sun on cold winter days. These foxes feed mainly on mice and insects. A dried-out springbok skull (*right*) lying alongside a well-worn game path on the barren surface of Etosha Pan bears mute testimony to an untold story of death in this harsh African landscape.

A pair of high-spirited lionesses, bellies bloated from the previous night's feast, gambols playfully alongside Goas waterhole (*above*). Lions are the only truly social cats, and interplay such as this forms an important bonding role among members of a pride, and may also serve to condition them for possible conflicts with rival prides or even marauding hyaenas. A redbilled hornbill (*right*) feeds her eager offspring. Etosha is a bird-watcher's delight, with some 340 species recorded on the official list. This includes three other hornbills, the grey, the yellowbilled and the rare Monteiro's hornbill.

Dust rises as springbok scatter before the determined charge of a lone lioness at Okaukuejo waterhole (*right*). 'The Old Girl', as she became known, took up semi-permanent residence at this waterhole overlooked by the Okaukuejo restcamp, and regularly lay in ambush for the thirsty herds which depended on this source of water. She killed on a regular basis, and provided excitement and interest for numerous visitors to the camp, before she herself fell victim to a farmer who laid out poison bait on his farm adjoining Etosha. The horned adder (*below*) is commonly encountered basking in the sun on roads and in other open places in late autumn, before the onset of the cold winter weather.

Giraffes are at their most vulnerable when drinking (*left*) and are generally very circumspect when approaching water, often spending several hours circling and inspecting their surrounds before bending to drink. They have special valves in the bloodvessels of the neck to regulate the flow of blood when they bend to drink, and when they rise with a sudden upright swing of the head while the forelegs are rapidly pulled together. A lappetfaced vulture, on its final approach, comes in to land (*above*). This species is one of the world's most impressive raptors, weighing about seven kilograms and with a wingspan approaching three metres.

Giraffes feed on thorny acacias with impunity. The 40-centimetre-long tongue twists around a thorny branch and pulls it into the mouth where it is held between the teeth of the lower jaw and a hard, bony pad in the upper jaw. Leaves and other edible parts are stripped from the branch with a backward jerk of the head, and any inedible pieces of the branch are discarded as they feed.

A black rhinoceros rolls a tsamma melon under its chin in an attempt to break the succulent fruit apart. These melons are relished by numerous animal and bird species, particularly in arid areas, and were prized by early Bushman hunter/gatherers as a valuable source of liquid sustenance. Unable to fit the whole melon in its mouth, the rhino eventually broke it with a foot before consuming it with apparent enjoyment.

The brightly coloured flowers of the sickle bush or Chinese lantern enliven the drab landscape in the spring months (*above*), although spectacular sunsets, ranging from singular molten balls of liquid gold (*left*) to brilliant kaleidoscopes of colour, are a year-round feature of the Park.

Curious but uncertain, a young lion watches intently (*left*) while we lie outstretched on the ground to achieve this low-angle photograph. After watching nervously for some time the lion relaxed noticeably, but the photo session was ended a short while later when a fully grown lioness approached to investigate more closely. A newly hatched ostrich chick (*below*) settles uncertainly in the face of the camera, too young to flee at the approach of potential danger. From the age of a few weeks, however, the chick will be capable of maintaining speeds in excess of 50 kilometres an hour. The chick featured here, only centimetres high, and with a mass of less than a kilogram, will grow to almost two metres when mature and will attain the mass of an adult man.

Salvadora, a contact spring on the fringe of Etosha Pan near Halali, is an important watering point for the large herds of game that throng the plains during the dry season, even though its water is extremely brackish. Here blue wildebeest, Burchell's zebra and a springbok herd drink together in the late morning heat, the shimmering Etosha Pan stretching far into the distance behind them.

A young leopard makes use of a temporary rainwater pool to satisfy its thirst. When rain fills holes such as this during the wet season, most animals abandon the more conventional drinking places to take advantage of the fresh, sweet water; most of Etosha's permanent water sources are high in saline content and are utilized only out of necessity during the dry winter months.

Bat-eared fox pups peer curiously from the safety of their den as the last rays of sunlight cast a golden glow over a lone springbok nearby (*above*). Soon they will emerge to join their parents in nocturnal foraging, feeding on termites and other subterranean insects which they locate by means of their large, sensitive ears. Although active at night, bat-eared foxes are frequently seen in the early mornings and late evenings sunning themselves in the vicinity of their den. The distinctive black lines running from the cheetah's eyes to the corner of its mouth (*right*) are the primary aid in distinguishing the species from the similarly spotted leopard. The cheetah's spots, however, are solid and isolated whereas those of the leopard are clustered in rosettes.

Pied crows (*above*) rarely fail to amuse with their raucous cries and attention-grabbing antics, though the tendency of some visitors to reward them with titbits is both illegal and destructive, for the birds may ultimately become dependent upon human handouts. Mountain zebras (*below*) enjoy a dust bath, an activity which performs a dual purpose, regulating body temperature and controlling parasites. These zebras also appear to enjoy mud baths to a far greater extent than their relatives, Burchell's or plains zebras, and are often seen caked with mud and clay.

A herd of Burchell's zebras nervously breaks from drinking as one of their group is startled by an imagined danger, perhaps nothing more than a bird passing overhead. Most prey species are easily alarmed and extremely timid when drinking, as this is when they are most vulnerable to sudden attack from predators such as lions.

The graphic stripes of zebras (*left*) are thought to play a dual role, serving to confuse attackers when the herd stampedes, and regulating body temperature. Honey badgers (*above*) are renowned for their ferocity and bravery, and several accounts are on record of them having killed creatures far larger than themselves, in most of these cases the badger having triumphed by seizing its adversary's scrotum. As their name suggests, however, honey badgers have a liking for honey and the larvae of wild bees, although they also eat reptiles, fruit, berries and the larvae of dung-beetles and other insects.

Masked weavers (*above*) are among the many colourful bird species that bejewel the Etosha countryside. Elephants (*right*) mill about in the midday heat, awaiting a turn at the water at the small Ngobib waterhole in the eastern sector of the Park. Elephant herds have a distinct social structure, and a complex hierarchy that strictly regulates which herd members drink when, particularly at a small water source such as this. Research undertaken in Etosha and elsewhere in Africa shows that elephants have a complex system of communication, using low-frequency rumbles to stay in contact with each other over a range of several kilometres.

To satisfy their giant thirsts, elephants must consume some 160 litres of water daily (*above*). This is accompanied by a daily intake of between 150 and 300 kilograms of dry food. Elephants browse and graze, a favoured browse being the new growth on the ubiquitous mopane tree. These animals can be highly destructive in their feeding habits, breaking or uprooting entire trees for the sake of a few mouthfuls, only to repeat the process at the next tree. They are gregarious and live in family units consisting of a matriarch or adult cow, who leads the herd, and her offspring of several generations, as well as a number of other closely related females and their progeny. Wildebeest (*right*) stand in silhouette against a winter sky, laden with dust and smoke from veld fires, and burnished by the setting sun.

ETOSHA: A WILDLIFE PORTFOLIO

Ostriches run across a flooded Etosha Pan (*left*), where they spent the night safe from predators, frightened at the sound of the microlight aircraft in which we were flying. Redbilled queleas (*above*) fly in a seemingly endless procession to the water throughout the day. Flocks of these birds can number in the hundreds of thousands, and have been known to obliterate entire grain crops in less than a day. In many parts of Africa they are regarded as an agricultural scourge, more reviled than locusts, and are relentlessly exterminated.

Sparring between giraffes (*left*) is believed to be related to the hierarchy within bachelor herds, though more serious duels between adult males take place over females in oestrus. The giraffes usually engage in these tussles by standing shoulder to shoulder or head to tail and swinging their heads and tufted horns into one another's sides; such 'fights', however, rarely lead to more than injury to the loser's dignity. Bath time for this blacksmith plover (*above*) means a chance to shed its mantle of dust. Despite its fearsome appearance, Bibron's gecko (*below*) is a harmless nocturnal creature that feeds primarily on insects.

Bull elephants, such as this one enjoying a sundowner (*above*), are usually solitary animals, though bachelor groups may be established on a temporary basis. Old bulls, however, are frequently accompanied by two or three younger male 'askaris' which act as body guards when the older animal's senses begin to fail. Males will only join up with a herd when a cow is in oestrus. Springbok (*right*) mass together, apparently apprehensive, staring fixedly at some unseen danger ahead. It is often possible to locate hidden predators by noting the behaviour of prey species, for once they've discerned the presence of an enemy they will generally stand in silence, staring in its direction. In this case the springbok have picked up the scent of a leopard lying at the roadside upwind of them; they eventually detoured around the waiting predator.

ETOSHA: A WILDLIFE PORTFOLIO

A young yellow mongoose greets its mother on her return from a foraging expedition (*left*), sniffing about her mouth in search of a tasty morsel. These predatory mammals eat beetles and their larvae, crickets, grasshoppers, termites and occasionally small birds. Young elephants (*above*) indulge in a playful sparring session in preparation for the day when they too may need to defend a herd, pushing and wrestling with their trunks and stirring up the dust about them. Codes of conduct in elephant society involve numerous greeting rituals and ceremonies.

The popular and ubiquitous helmeted guineafowl (*above*); its raucous squabbling is one of the characteristic sounds of the African bushveld. A pack of wild dogs (*right*) moves off to begin the hunt. These unusually marked animals, whose name translates from the Latin *Lycaon pictus* as 'painted wolves', are seriously endangered throughout their range. They have a highly developed social infrastructure and hunt co-operatively in packs, killing and devouring their prey on the run, a technique that caused them to be reviled and persecuted by hunters and even by many game wardens until recent years. Widely misunderstood, they rank as the most efficient hunters of the African savanna with a success rate far greater than any other predator. Namibian wildlife authorities are making a concerted effort to increase the numbers of this species in Etosha.

A flock of ostriches (*above*) moves on to the edge of Etosha Pan to drink from the Okondeka seepage. Once satisfied they will make towards the centre of the open pan where they will spend the night, safe from the attentions of predators. Ostriches also enjoy grazing on the nutrient-rich, short grasses that grow on the pan, and can often be seen far out on its surface, as if in the middle of nowhere. Although less adept at tree climbing than leopards, lions do readily climb where suitable trees are present, and also seem to enjoy clambering about fallen trees (*right*) and other raised objects, which often serve as vantage points too.

The cryptic colouring of a tiny scops owl (*left*) ensures that it is perfectly camouflaged where it sleeps in the crook of a branch in a mopane tree. One of the smallest owls, its chirruping trill is a common sound in the restcamps at night. A pair of elephant bulls (*above*) continues feeding in a patch of acacia thorn scrub, oblivious of the early summer bush fire which rages across the landscape behind them. Although most animals manage to move clear of such fires, often started by park authorities as part of the veld management programme, larger and slower species such as elephant and rhinoceros do occasionally get caught and seriously burned.

Rainclouds gather in the western sky as another day draws to a close around the still waters of Two Palms waterhole (*below*). A majestic flier with a rocking flight, the bateleur (*right*) is one of the most magnificent and colourful of the eagles. Its common name is said to refer to its habit of somersaulting and performing other acrobatics in the air. These raptors feed on snakes, lizards, small mammals and carrion.

Reputed cowards and skulking scavengers, spotted hyaenas (*left*) are in fact major predators and a large proportion of their food requirements derives from prey which they have hunted themselves. Hyaenas compete seriously with lions and lesser predators, and will readily challenge and drive lions off their kills, particularly in areas where the hyaena population is large; conversely, though, it has been found that lions frequently scavenge hyaena kills, reversing the popular images of these two species. Soaked through after an afternoon rainstorm, a herd of springbok grazes contentedly (*above*), their white dorsal crests raised to dry in the evening breeze.

Dainty and graceful, a steenbok ram (*right*) browses among acacia thorn and other scrub. Generally solitary animals, the males do join up with females in oestrus, and females may be seen in company with their offspring. Only the males have horns, though horned females do occasionally occur. Their common name derives from the Afrikaans word for a brick, *steen*, which describes the reddish colour of their coats.

Common throughout Etosha, particularly in the restcamps, yellowbilled hornbills (*above*) readily adapt to human proximity and will brazenly invade a picnic table to snatch any morsel that catches their eye. Otherwise their diet consists predominantly of insects, fruit and small lizards. The redcrested korhaan (*right*) is renowned for its spectacular aerial courtship display, flying straight up to some height, then suddenly closing its wings and plummeting earthwards as if certain to crash into the ground, only to open its wings and glide gracefully for a short way before settling.

A lone wildebeest grazes in a field of yellow, daisy-like *Hirpicium gazanioides* flowers (*above*), while several marabou storks prowl behind, and a large herd of springbok feeds in the distance. Floral spectacles such as this appear briefly after the onset of the summer rains. A leopard (*right*) concentrates its attention on a small herd of impala passing in the distance.

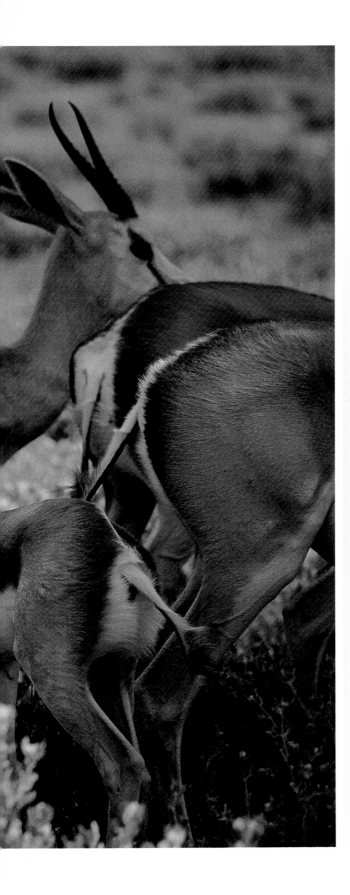

Summer is a time of rejuvenation, and with the arrival of the rains comes the lambing season when an atmosphere of expectancy and excitement pervades the grasslands. The springbok (*left*) are among the first to begin dropping their young, and within days of the early rains scores of lambs can be seen frolicking on the plains. Lambing is timed in such a way that ewes in a herd drop their young within a short period of one another, thus giving the newborn a greater chance of survival. Springbok lambs are extremely vulnerable in the first few days of life, and are usually hidden among tufts of grass or under low bushes until they gain in strength and can take their place in the herd. During this period they frequently fall victim to predators such as leopards (*above*), and even to large raptors.

Yet another fascinating facet of the wildlife in Etosha, a yellow and black blister beetle (*left*), here seen feeding in a *Hirpicium gazanioides* flower. The crimson-breasted shrike (*below*) was formerly regarded – unofficially – as the national bird of this once German colony, its striking red, black and white colouring reminiscent of the German flag. The diminutive Damara dik-dik (*right*) is Etosha's smallest antelope, attaining a height of about 40 centimetres and a weight of less than five kilograms. Dik-diks have characteristic proboscis-like snouts, which can be extended in any direction to sniff at food. These animals are primarily browsers, favouring small shrubs, but will also graze on fresh grass shoots during late spring and early summer. They are restricted to the eastern part of Etosha, and occur singly or in family groups comprising a male, a female and a single young.

Life and death on Etosha's plains. A springbok ewe licks tentatively at her emerging lamb (*left*) which remained protruding like this for more than an hour, during which time the ewe continued with her normal routine. Later she lay down (*above*), racked in the throes of labour while a ram stood attentively nearby, apparently on the lookout for marauders. A ewe is extremely vulnerable to attack at times such as this, but in this instance was protected by a concerned and mindful ram. Following more than two hours of labour the lamb was eventually thrust free, after which the ewe roughly licked it dry before encouraging it to rise on its unsteady legs (*right*). After a stressful birth, the ill-fated lamb fell prey to a lone cheetah later in the day (*far right*), which stumbled across it lying hidden among grass tussocks.

African bullfrogs, the size of dinner plates (*left*), emerge only after good rains have fallen, having remained cocooned underground in aestivation since the rains of the previous summer. Their sudden appearance is marked by loud croaks reverberating across the countryside, and vicious territorial fights as males battle for possession of the choicest breeding ponds. The frogs are aggressive and possess two strong tooth-like protuberances in the lower jaw, which can inflict nasty wounds. After gaining possession of a pond, the male is joined by several smaller females, and mating ensues. Etosha elephants (*above*) are noted for their small, stunted, and often broken tusks, the result of mineral deficiencies in their diet.

Although leopards can go without water for extended periods, obtaining all their liquid requirements from their food intake, they will drink regularly when water is readily available. Showing a preference for fresh rainwater, this leopard (*above*) prepares to drink from a muddy roadside puddle rather than from the large but brackish fountain nearby. Both lesser and greater flamingoes turn the shallow waters of Etosha Pan into a spectacle of vivid pinks, reds and whites (*right*).

A herd of elephants increases pace as it nears Gobaub waterhole, a remote fountain not accessible to the general public as yet. Scattered heaps of dung and a broad, well-worn path bear testimony to the large number of elephants that utilize this drinking place. Elephants appear to tread the same paths to and from their regular feeding and drinking places, wearing smooth, well-defined trails that are used by subsequent generations. The large cow in the left of the picture is the matriarch, or leader of the herd.

A small and sinuous slender mongoose (*above*) interrupts its early morning foraging for insects, mice and lizards to peer curiously at the camera. Elephant calves (*right*) may be born at any time of the year, after a 22-month gestation period. They weigh about 120 kilograms at birth, and are pinkish in colour with a generous covering of hair on their bodies. Young calves have poor eyesight and rarely move more than a few metres from their mothers, maintaining contact by touching with their rubbery trunks. They suckle for up to three years, although they start eating leaves and twigs when just a few months old. Maternal protection of the young is vigorous, and breeding herds of elephants are notorious for their aggression towards intruders. The young calf walks under its mother's belly when on the move, where it is well protected against enemies.

Eland (*left*) are Africa's largest antelope, weighing as much as 800 kilograms and attaining a height at the shoulder of almost 1,8 metres. Both the male and female carry horns, but those of the male are much heavier than the female's. Although occurring in small herds throughout Etosha, they are not commonly seen other than in the eastern sector of the Park near Namutoni, where they browse among the terminalia and combretum woodlands. When walking, eland produce a curious clicking sound, believed to come from their knee joints. One of the larger birds of prey, the tawny eagle (*right*) is common throughout most of southern Africa. As seen here it often perches in dead trees, on the lookout for prey such as small mammals, snakes, lizards and large game birds, although it will also eat carrion where available. A young Cape fox (*below*) sniffs curiously at the ground, perhaps on the trail of a mouse or tasty insect that passed earlier.

Cheetah cubs remain with their mothers until maturity and family groups such as this (*left*) are regularly encountered. Although less playful than lion cubs, perhaps because of their asocial family structure, cheetahs are innately curious, the three juveniles here exploring the saline formations at the Suaeda spring on the road between Okaukuejo and Halali. The fork-tailed drongo (*above*), here seen stretching its wing in the late afternoon sun, is one of the more common birds of the bushveld.

In years of good rainfall flamingoes flock from across Africa to feed, and if conditions are favourable to breed in the nutrient-rich waters that flood Etosha Pan in summer. As these shallow waters begin to evaporate, spectacular flocks of the pink birds can be seen flying across the Pan's expanse (*above*) in search of any remaining water. In some years the absence of late rains means that newborn chicks are left stranded, starving on rapidly drying mudflats, where they fall victim to marauding predators such as jackals and hyaenas. Park authorities do all they can to rescue the chicks, but often thousands perish. After a dry and dusty trek through the mopane woodland a pair of elephants (*right*) enjoys a refreshing drink and cooling mudshower at Agab waterhole in the central part of the Park.

Bat-eared foxes (*left*) rest alongside a termite mound they may well have broken open in search of a tasty feast. The foxes use their large, highly sensitive ears to listen for subterranean sounds made by insects such as termites, which they then dig out using their strong front claws. The unique and seriously endangered desert black rhinoceros of the subspecies *Diceros bicornis bicornis* (*below*) has adapted to life in the harsh, dry environment of the Kaokoveld. Comprising Damaraland and Kaokoland, this starkly beautiful country to the west of the Park is rich in wildlife. Other large mammals here include elephant, giraffe and also gemsbok, springbok and mountain zebra. This area was originally proclaimed as part of Etosha in 1907, but subsequent deproclamations saw it returned to tribal ownership.

Seasonal migrations of wildebeest occur once the summer rains are in full spate, though the counter-clockwise movement that took place in previous years no longer occurs because of the positioning of Etosha's northern boundary fence. Today the move is an east-west one, wildebeest and other large herbivores trekking to the usually arid Grootvlakte area west of Okaukuejo which offers new pastures and a supply of fresh water. However, as the dry season approaches and the water evaporates, the movement will reverse and the animals will move east once again, retracing their route of a few months earlier. This means that these lands are utilized twice in a season rather than once, as was the case before human interference, a serious cause for concern for Park biologists.

Graceful and stately, the kudu bull (*left*), with his long spiral horns and delicate markings, is one of the most spectacular animals of Africa. Though shy and timid by nature, the kudu can put its rapier-sharp horns to effective use in defence, and predators frequently come off second best in an incautious attack. Awaiting its turn at Okondeka seepage, a giraffe (*above*) watches while a gemsbok bull mounts another. Such behaviour is believed to be part of a strict dominance display by territorial males over other males within their range. Courtship and mating are the prerogative of the territorial male and other males are tolerated within his territory so long as they submit to the hierarchy of absolute dominance upon which the herd is structured. Threat-displays and horn clashing between males also form part of the territorial imperative.

Expressing his anger at the intrusion, this lion (*above*) circled our flimsy tent several times in our camp in a remote part of Etosha. Lying awake at night we could hear the soft pad of the big cat's feet in the dust as it prowled outside, and a low throaty growl as it investigated our scent. Etosha lions have a fearsome reputation, and are allegedly larger than most of those found elsewhere in Africa. The spoor of this lion imprinted heavily in the soft soil near our camp (*right*). Secretary birds (*far right*) are found in pairs throughout the year. A good deal of their daily activity centres around the roost, usually on top of a large, flat-topped tree. Occurring throughout Africa south of the Sahara, in Etosha they are commonly encountered striding and strutting across the grasslands in search of their prey, which can be virtually anything from grasshoppers to rats, mice, small birds and even deadly poisonous snakes such as puff adders and cobras. Their legs are covered with tough, thick scales impregnable to a snake's fangs. This bird is said to get its name from the long feathers protruding from the back of the head, these resembling the quill pens of old, though a more likely source is the Arabic word for a hunting bird, *saqr-et-tair*.

This colourful acraea (*above*) is one of many butterflies found in Etosha. By concentrating on less obvious species such as this, visitors to this unspoiled wilderness can greatly enhance their experience of the bushveld, than if they merely seek out the big game for which the Park is justifiably famous. Imposing and stately, elephants (*right*) assemble at the floodlit waterhole in the Okaukuejo restcamp every evening during the dry season, where they provide entertainment for fascinated residents well into the night.

Making certain that not a scrap of meat remains, three lion cubs (*left*) chew on the left-overs of a springbok killed earlier by another pride member. Reintroduced to Etosha a number of years ago, and afforded special protection in the endangered species' enclave of Kaross in the western corner of the Park, roan antelope (*above*) appear to be breeding well. Several groups have now been released in the central part of the reserve where they are regularly seen near the m'Bari waterhole. Fully grown, roan are second in size only to eland.

Storming from the cover of a low-hanging shrub where she has waited in ambush, a lioness startles a young kudu drinking deep within a waterhole. The kudu is unable to flee and is soon overtaken by the fleet-footed predator which throws her weight on to the frail young antelope's back. Dust flies as the two roll with the impact, before the lioness locks her jaws around the struggling kudu's throat. After several weakening kicks the struggle is over and the successful huntress straddles her victim, before dragging it back to the shade of her lair. Moments later the kudu herd returned to the waterhole, secure in the knowledge that, having sacrificed one of their members, they were free to drink in safety.

A black-backed jackal (*left*) scavenges the remains of a cheetah kill. This opportunistic canid is widespread throughout Etosha and is common inside the restcamps at night, where it has been known to snatch sizzling meat from barbecue fires before the eyes of startled visitors. Namaqua sandgrouse (*above*), here seen coming in to land among a herd of springbok, flock to water at a regular time each day, usually one to three hours after sunrise, arriving in noisy flights that may number in the thousands. An inhabitant of the drier regions, this species is noted for the male's habit of providing his young with water by soaking his breast and belly feathers.

Seemingly burdened by weighty problems, zebras commonly adopt such head-on-back poses (*above*), standing lethargically in small clusters, each animal resting its head on another. Although appearing identical at first glance, no two zebras are exactly alike, their stripes having the same uniqueness as human fingerprints. The same can be said of the blotches of the curiously marked wild dog (*right*); no two of these animals have the same or similar markings. This dissimilarity aids researchers in identifying individuals within groups, particularly useful at present as this species is endangered and is currently the subject of much study.

A young leopard (*above*) makes a meal of a terrapin it has hooked out of a muddy rainwater pool. Though quite able to tackle large antelope, leopards are opportunistic killers and will readily consume prey such as this if it is available. Primarily arboreal, chameleons move to solid ground (*right*) when it is time to breed, travelling widely in search of a mate. Although masters of camouflage, claims of their ability to change colour to match their surroundings have been highly exaggerated. Chameleons generally have the same base colour as their habitat, and change colour more in response to light and temperature than to their background environment.

The rangers at Etosha have devised a unique method of monitoring and compiling identification dossiers on the Park's black rhino population. By staking out waterholes during full-moon periods at night – when rhinos most commonly drink – and photographing the usually aggressive but shy animals as they come down to the water (*left*), authorities have succeeded in compiling accurate identikits of more than 300 of the Park's estimated 340 black rhinos. This technique is not as easy as it might sound, as the photographer is required to quietly stalk towards the drinking animal, taking photographs from range markers as close as 15 or 20 metres.
A family of zebras (*above*) continues feeding, unmoved by the spectacular sunset behind them.

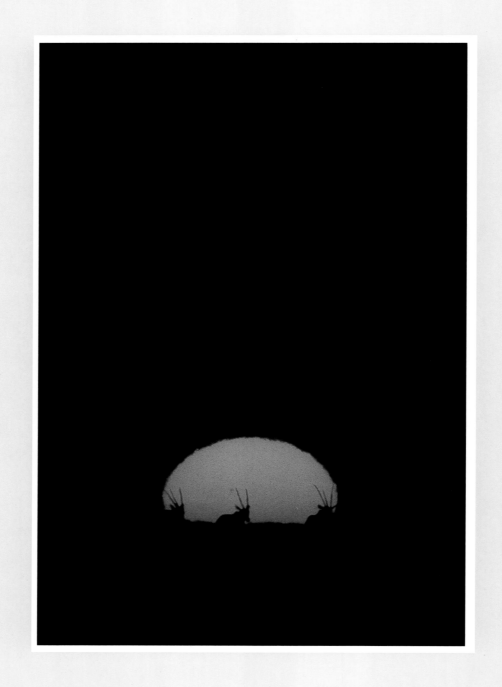

PHOTOGRAPHERS' NOTES

Etosha is a wildlife photographer's dream, and a commission to produce a book on the Park is undoubtedly a dream assignment. The opportunity to live for almost a year in one of Africa's greatest wildlife refuges will linger in our memories forever.

Wildlife photography is for us totally absorbing, and always presents a new challenge. It has its pitfalls and drawbacks, but these are far outweighed by the satisfaction and pleasures we enjoy. For eight months of every year we live under canvas, sleep on a mattress on the ground, cook on gas or wood fires, and forego the day-to-day conveniences most people take for granted. Weeks may go by without fresh provisions, or news from home, or the unimaginable luxury of a cube of ice in a glass of whisky at the end of a dry and dusty day. We weigh these minor hardships against a fresh, crisp sunrise spent with a pride of lions or a family of squirrels, a lunch-time parked in a minimum of shade alongside a busy fountain, entertained by herds of elephant splashing and bathing in happy abandon, and falling asleep to the whoops and moans of a prowling hyaena, the yelps and yips of a lonesome jackal and the distant roar of a mighty lion.

We prefer to work on foot and away from the beaten track whenever possible, and have been fortunate in gaining permission from the relevant authorities to do just that. Although many of the photographs appearing in this book have been taken at places inaccessible to the general public, this was to afford ourselves the privacy of working undisturbed and with animals that were not distracted by other people or vehicles, rather than because they cannot be seen in other areas. Apart from game such as mountain zebra, klipspringer, baboon and perhaps roan antelope, which occur predominantly in the western part of the Park, all the animals featured can be seen quite readily throughout Etosha, particularly at the many waterholes.

As far as camera equipment is concerned, we have always used Minolta X700 cameras and have come to rely on their trouble-free operation under harsh conditions. Photographing wildlife often calls for 'point-and-shoot' action, and the built-in light-meters of these cameras are entirely dependable in even the most difficult of conditions. Light in the African bushveld can be harsh one moment and extremely subtle the next, and accurate metering is essential. When time allows – and most often it does not – a Minolta Spotmeter F makes a valuable accessory.

Our favourite lens is a 500 mm f4,5. This offers ample magnification and is fast in low light. It is far easier to handle than an equivalent 600 mm, and by sacrificing 100 mm we save ourselves several kilograms in weight and gain a lot more mobility. A Minolta 2x converter used in conjunction with the 500 mm gives us a 1 000 mm f9, and provides us with the extra reach should we need it. We supplement these with a 300 mm, an 80-200 mm f2,8 which we find superb in low light and in conjunction with a flash for night work, a 90 mm Macro, a standard 50 mm, and a 28 mm, 24 mm and 17 mm for wider angle shots. For night photography we use Metz 60 CT-4 and 45 CT-1 flashes, all with 'Televorsatz' flash intensifiers for extra range. A recent addition to our battery of manual cameras is an autofocus Minolta Maxxum 7000i with a 100-300 mm zoom lens. Its combination of super-quick and predictive autofocusing with compactness and light weight makes following action fairly straightforward, and its built-in spotmeter makes calculating tricky exposures easy.

All of this equipment is carried in our most indispensable accessory – an airtight, dustproof, waterproof and unbreakable Pelican case. Expensive and difficult to obtain in South Africa, these are essential for the outdoor action photographer in harsh climates.

For ease of access while working from our vehicle, we store our equipment in a cloth bag sewn with several pockets, which hangs between us over the seat. In it, ready for action, we store four cameras with different lenses. This allows us to reach for a camera and operate it immediately, avoiding fumbling with lenses at a critical moment.

We prefer using Kodachrome 64 film, despite the drawback of having to send it abroad for processing. However, we have experimented with alternatives and have found the new Fuji Velvia 50 to be superior in every respect – sharpness, grain, colour fidelity and saturation – to the old Fujichromes, and it can be processed locally. All our E6 processing is handled by Creative Colour in Cape Town, under the personal supervision of Dennis Sprong.

We avoid using tripods whenever possible, and have tried numerous door and window-mounting gadgets for the use of telephoto lenses from the vehicle. All have been discarded in favour of a simple steel table which fits over the window ledge and upon which we place large beanbags to steady our cameras. Likewise, when shooting away from our vehicle, we place our cameras on beanbags, flat on the ground. And a steel plate, threaded to fit on top of a sturdy set of Manfrotto tripod legs, makes a far steadier stand when used with beanbags than even the most expensive of tripod heads.

Our preferred angle of photography is from below the eye level of the subject, and whenever possible we will leave the vehicle and lie on the ground to get the lowest possible perspective. This is not without its dangers, however, and should not be attempted by the inexperienced. It is also illegal in most game reserves without special permission. However, it is not always possible to get out of the vehicle or work on foot, and in these cases we must shoot from the vehicle windows. Because of this we have always used Isuzu 4 × 4 pick-up vehicles, as they offer adequate ground clearance without the exaggeratedly high bodies of most other off-road vehicles, and so allow for the lowest possible angle of view in such circumstances. They are also comfortable, economical, rugged and reliable – all crucial attributes.

Whenever we set out from our camp in the morning we know that we may be gone all day, often longer, and for this reason always maintain a 'lug-box', complete with a small gas cooker, kettle and pan, and basic necessities, which never leaves the vehicle. A good supply of fresh water is essential in the bushveld, and we always carry extra fuel, a second fully charged battery and two spare tyres, as well as a puncture repair kit and basic vehicle spares. A high-lift jack completes the stock of equipment, this being a tool we have learnt the hard way should never be left behind when venturing off the beaten track with a 4 × 4 vehicle.

TRAVELLERS' INFORMATION

GETTING THERE

Etosha is situated in the far north of Namibia, 435 kilometres on good, tarred road from Windhoek if your destination is the main camp of Okaukuejo, and 535 kilometres from Windhoek if you are heading for Namutoni.

Those driving to the Park should take the B1 highway north from Windhoek, branching off on to the C38 at Otjiwarongo and making for Okaukuejo via Outjo; or, if Namutoni is where you are heading, continue straight on the B1 to Tsumeb and follow this road towards Ondangwa, turning left towards Namutoni (as signposted) about 72 kilometres from Tsumeb. The roads are well signposted and well maintained, and travellers should have no problems in this regard.

Windhoek, the capital of Namibia, is a modern city with all conveniences. It boasts two airports, and the national airline, Namib Air, offers a comprehensive domestic service using 19-seater turboprop aircraft, as well as regular flights to and from Johannesburg and Cape Town in Boeing 737s. There are also international flights twice a week, between Windhoek and Frankfurt, on spacious 747 jumbo-jets, as well as numerous other flights which link the capital with Botswana, Zimbabwe, Zambia and Kenya.

The commercial airport nearest to Etosha National Park is at Tsumeb, 107 kilometres from Namutoni, and is served by daily scheduled flights. There are car rental agencies throughout the country, including Avis and Kessler, which offer a range of sedans but specialize in fully-equipped 4 × 4 vehicles.

All three of the restcamps, Okaukuejo, Halali and Namutoni, maintain gravel airstrips which may be used by private and commercial charter flights, and there is also an excellent all-weather strip at Mokuti Lodge. Before using these strips, pilots should check with the relevant authorities – in the case of the restcamps with the tourist office at the camp concerned. Pilots must circle the restcamp before landing in order that transport may be sent for the collection of passengers. No aviation fuel is available within the Park.

GETTING AROUND

An alternative to driving yourself to and around the Park is to join a guided tour. There are several operators offering safaris to Etosha, from small, exclusive operations such as Bushdrifters, Etosha Fly-In Safaris and Namib Wilderness Safaris, to big bus-tour groups like Suidwes Safaris and Oryx Tours. These safaris are often preferable for people visiting Africa for the first time, as the drivers are all experienced guides conversant with the flora and fauna of the Park. However, the tourist roads in Etosha are gravel surface and well signposted, so there is no reason why even the most inexperienced should not hire a car and drive around the Park, this being the most intimate way to enjoy the wilderness. A comprehensive map of the Park's road network is available from the shops in the restcamps, along with other pertinent and useful information for visitors.

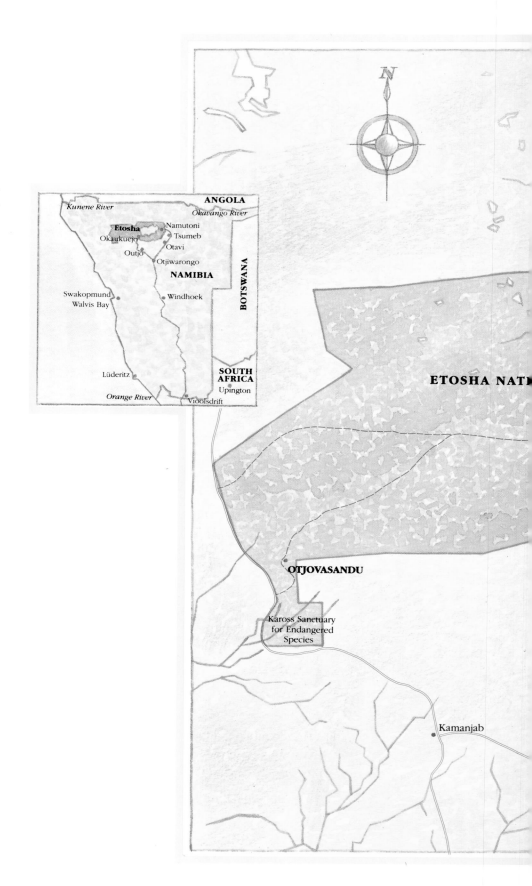

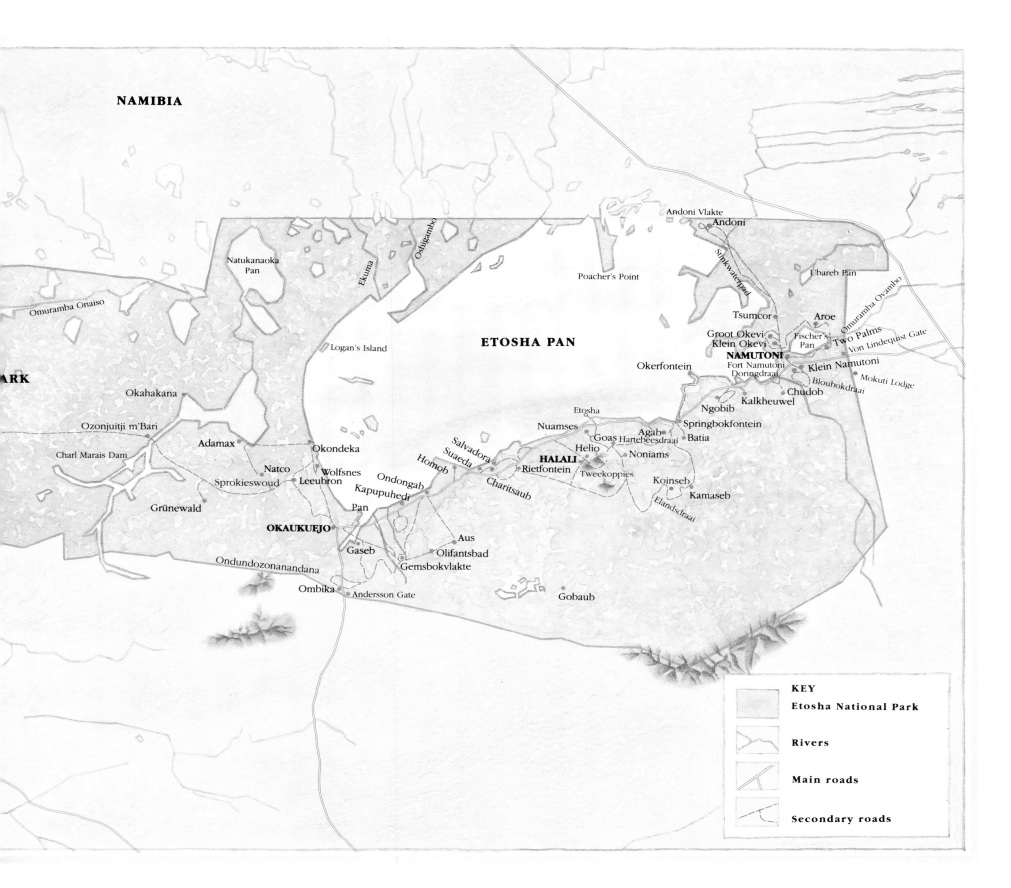

There are filling stations which sell petrol, diesel and oil in each of the three camps. They usually stock a limited range of spare tyres, and will undertake puncture repairs, but there are no other vehicle repair facilities. Gas bottles may also be filled.

Open vehicles and motor-cycles are prohibited inside the Park, as are pets, airguns and unsealed firearms.

ACCOMMODATION

Accommodation within the Park is provided by the Namibian conservation authorities in three restcamps, while plans to open a fourth in the west of Etosha near Otjovasandu are being considered. It comprises comfortable two-, three- and four-bedded bungalows, two-bedded 'bus-quarters' or motel-rooms, four-bedded mobile homes (Namutoni only) and four-bedded tents. The formal accommodation at Namutoni and Okaukuejo is air-conditioned. In addition there are camping and caravan sites at all three restcamps, many of them serviced with on-site electric power-points. There is a swimming pool in each of the restcamps.

The Okaukuejo and Namutoni restcamps remain open all year round, but Halali closes during the hottest months from 1 November to 15 March.

All of the camps have fully-licensed à la carte restaurants serving breakfast, lunch and dinner daily, as well as well-stocked shops which sell fresh meat, eggs, cheese, butter, cold drinks and ice, as well as beer, wine and spirits. There are also tinned foods, toiletries and limited fresh vegetables and bread. We suggest stocking up with fresh vegetables at one of the larger towns en route, and buying fresh bread at Outjo or Tsumeb, however, as supplies in the camp shops usually do not last long and are only received once or twice a week. The shops also stock a range of gifts and curios, as well as a selection of books and field guides, many of which are available in German.

Hours are curious, the shops opening and closing at odd times throughout the day, and it is best to acquaint oneself with these on arrival at each of the camps. The restaurants also have limited hours, although there are snack kiosks in all three camps which open for periods outside of the restaurants' open times.

Visa and Master Card credit cards as well as traveller's cheques and bank guaranteed cheques in local currency are accepted at all camps. It is best to check with the Automobile Association or your bank, however, as to the acceptance of various petrol and garage cards, for several of these are no longer valid in Namibia.

Alternative accommodation to that within the Park is available at the luxurious Mokuti Lodge, a Namib-Sun hotel situated 500 metres outside the Von Lindequist Gate, near Namutoni. The Lodge offers daily game drives into the Park, as well as night drives within its own area, which has a wide variety of game. Transfers to and from the airport at Tsumeb can be arranged, and vehicles can be hired at the depot there.

For those eager to see the wilder country to the west of Etosha, the privately run Hobatere Lodge is situated in spectacular countryside that once formed a large part of the Etosha National Park. The Lodge is sited among rugged dolomite kopjes in outstanding game country, where mountain zebras, klipspringers and other species not likely to be seen by visitors to Etosha, may be encountered. Lions, elephants and other big game may also be seen here, on privately conducted game drives or walking safaris.

WHAT TO TAKE

The climate varies considerably from winter to summer. Winter nights can be very cold although the days are pleasantly mild. A warm jersey, jacket or wind-breaker is essential for the early mornings and the nights, particularly if one intends sitting up at the Okaukuejo waterhole. The average daily minimum for July is 6 °C. Summers, however, can be rainy and are generally very hot, December averaging a daily maximum of about 35 °C. A good sunscreen, lip balm and a broad-brimmed hat are useful items. Even the mid-winter sun in Namibia can cause severe sunburn.

A pair of binoculars is an essential item for the full enjoyment of any safari. Any model in the 7×35 or 8×40 range is suitable for game viewing, although avid birdwatchers may prefer a more powerful combination of 10×40 which will mean extra weight, however.

No trip to a game reserve is complete without a field guide to birds, and also perhaps a guide to the larger mammals. There are any number of good natural history guides on the market today, and serious enthusiasts may wish to add to these, illustrated guides to the reptiles, insects, butterflies, trees and flowers. Indeed, learning to identify the trees and shrubs in a park can be as much fun as studying the birds.

Those wishing to try their hand at photography must ensure that they take an adequate supply of film, for it is not always available in the Park's shops. A good rule is to calculate how much film you *think* you'll need, and then double that figure.

The standard of your equipment and accessories will depend on your budget. A good 35 mm single lens reflex (SLR) camera, manual or automatic, is your basic requirement. It is ideal to have a selection of interchangeable lenses too, although one good telephoto will suffice. When photographing animals in the field, a 300 mm lens should be considered the minimum focal length. One of the new-generation zoom lenses in the range 75-300 mm or 100-300 mm is ideal. A wide-angle zoom, such as a 28-80 mm, makes an excellent companion to the above, providing a complete range of angles and magnifications from wide-angle to 300 mm telephoto in only two lenses. To extend the focal length of a lens a 1,4x or 2x converter makes a useful accessory, but means sacrificing quality to a certain degree.

Remember when buying equipment that it is better to spend more money on the lens than on the actual camera body, for it is the optics that determine the quality of the photograph.

Longer telephoto lenses in the 400-600 mm range bring you into a more serious category, and represent a sizeable investment. They also mean a greater degree of difficulty in use, for the higher the magnification, the steadier the camera/lens must be in order to attain acceptably sharp results.

A beanbag makes the perfect rest for the camera; it is easy to make and will ensure greater sharpness of image. A bank bag or similar item, half-filled with rice or small beans, is ideal. Air travellers with weight problems could simply take along an empty bag, a needle and thread, and then buy the rice in one of the camp shops and sew the bag closed after filling it.

The choice of film is personal and will depend on what kind of photograph you want to take. Modern colour print films give excellent results, even those with high ISO (or ASA) ratings show minimal graininess and offer considerable leeway in exposure. Slide or transparency films are far less forgiving, and exposures need to be very accurate. Most professional photographers use slide film and prefer an ISO rating of 50 or 64, as this film gives a fine, sharp image with minimal grain – essential for reproduction purposes. Generally a 100 ISO film is more than adequate for use during the day, with a few rolls of 400 ISO for low light conditions. Remember to keep cameras and films cool – exposure to heat will quickly ruin them.

SUGGESTED READING

Newman's Birds of Southern Africa; (Southern Books): The most popular and easiest to use of the field guides.

Roberts' Birds of Southern Africa; (John Voelcker Bird Book Fund): The birder's bible. The best reference work for more in-depth information, its format is not ideal for field identification.

Chris and Tilde Stuart's *Field Guide to the Mammals of Southern Africa*; (Struik): A well-laid-out photographic guide to big and small mammals, and with a great deal of interesting information.

Bill Branch's *Field Guide to the Snakes and Reptiles of Southern Africa*; (Struik): Another in the popular Struik series of photographic guides. Recommended for those with an interest in the smaller creatures of Etosha; the Park is home to 110 reptile species.

Trees and Shrubs of the Etosha National Park by Cornelia Berry. Available in the camp shops, this informative and easy-to-use guide was compiled by the wife of former chief biologist at Etosha, Hu Berry, and is illustrated by the now well-known black rhino conservationist, Blythe Loutit.

Trees of Southern Africa by Keith Coates Palgrave; (Struik): A comprehensive guide to the trees of the sub-continent. Big and bulky, this is more of a reference work than a field guide.

Lake Ngami by Charles John Andersson; (Various reprint editions): An historic account of travel by the first European to see Etosha. A true insight into Africa as it was before the advance and ravages of 'civilization'.

On Wildlife 'Conservation' by Ron Thompson; (United Publishers International, New York): A controversial but coherent look at conservation, its problems, and potential solutions in today's world.

Death in the Long Grass by Peter Hathaway Capstick; (Methuen Paperbacks Ltd, London): Absorbing tales of terror involving the denizens of the wild, it will keep you from stepping out of your car at the wrong moment.

The Besieged Desert by Mitch Reardon; (Collins): A graphic account of war, drought and poaching in pre-independent Namibia. A true account of the problems faced by wildlife conservationists in this arid land.

REFERENCES

ANDERSSON, C. J. 1856. *Lake Ngami*. Hurst & Blackett, London.

BALFOUR, D. and S. 1991. *Rhino*. Struik, Cape Town.

BANNISTER, A. and JOHNSON, P. 1978. *Namibia: Africa's Harsh Paradise*. Struik, Cape Town.

BERRY, C. *Trees and Shrubs of the Etosha National Park*. (Booklet available from shops in Etosha's restcamps)

BERRY, Dr H. 1989. *Etosha National Park*. Struik, Cape Town.

CHADWICK, D.H. 1983. 'Etosha: Namibia's Kingdom of Animals', *National Geographic* vol. 163, no.3, p.343. National Geographic Society, Washington.

CHANNING, A. J., DU PREEZ, L. and KOK, D. J. 1989. *Journal of Herpetology*. Society for the Study of Amphibians and Reptiles, Ohio University, Ohio.

CILLIERS, A. 1989. 'Monitoring methods and techniques for censusing black rhinoceros *Diceros bicornis bicornis* in Etosha National Park.' *Koedoe* vol.32, no.2, p.49. Pretoria.

COATES PALGRAVE, K. 1977. *Trees of Southern Africa*. Struik, Cape Town.

DU PREEZ, J. S. and GROBLER, I. D. 1974. *Drinking times and behaviour at waterholes of some game species in the Etosha National Park*. (Progress Report 1974) Division of Nature Conservation and Tourism, SWA. Windhoek.

EBEDES, H. 1976. 'Anthrax epizootics in Etosha National Park', *Madoqua* vol.10, no.2, p.99, Ministry of Wildlife, Conservation and Tourism, Windhoek.

JOOSTE, Prof. J. P. 1974. *Gedenkboek van die Dorslandtrek*. University of Potchefstroom, Potchefstroom.

JOUBERT, E. 1980. 'Mammal Adaptations for Desert Survival', *Rössing* December 1980, p.3-6, Rössing Uranium, Windhoek.

Journal of Gerald McKiernan in South West Africa: 1874-1879. Van Riebeek Society, Cape Town.

JURGENS, J. D. 1979. 'The Aura of the Etosha National Park', *Madoqua* vol.11, no.3, pp.185-208, Ministry of Wildlife, Conservation and Tourism, Windhoek.

LAMBRECHTS, H. 1985. *Namibia: A Thirstland Wilderness*. Struik, Cape Town.

McBRIDE, B. 1990. *Liontide*. Jonathan Ball, Johannesburg.

REARDON, M. and M. 1981. *Etosha: Life and Death on an African Plain*. Struik, Cape Town.

SELOUS, F. C. 1985. *A Hunter's Wanderings in Africa*. Galago Publishing, Alberton.

VEDDER, H. 1966. (translated and edited by Cyril G. Hall) *South West Africa in Early Times*. Frank Cass, London.

IMPORTANT ADDRESSES & TELEPHONE NUMBERS

Ministry of Wildlife, Conservation & Tourism Reservations Private Bag 13267 Windhoek NAMIBIA Telephone: (061) 36975 (Bookings) (061) 33875 (Information) Telex: 0908-3180	Bushdrifters PO Box 785743 Sandton 2146 SOUTH AFRICA Telephone: (011) 659-1551/2 Telex: 4-30753 SA Telefax: (011) 659-1122	Mokuti Lodge Namib-Sun Hotels PO Box 403 Tsumeb NAMIBIA Reservations Telephone: (061) 33145 or direct (0671) 3084 Telex: 0908-3138 or direct 0908-749 Telefax: (0671) 3084	Hobatere Lodge PO Box 110 Kamanjab 9000 NAMIBIA Telephone: 0020 ask for Kamanjab 2022	Namib Air PO Box 731 Windhoek NAMIBIA Telephone: (061) 38220	Avis Car Hire PO Box 2057 Windhoek NAMIBIA Telephone: (061) 33166	Kessler Car Hire PO Box 20274 Windhoek NAMIBIA Telephone: (061) 33451 Telefax: (061) 224551

INDEX

Page references in *italics* refer to photographs

aardvark 36
aardwolf 36
Acacia nebrownii 13
 A. nilotica 13
acraea *156*
adder, horned *84*
Agab *12*, 19, 28, *74*, *146*
Alberts, Gert 10
Aloe littoralis 13
Amutoni 11
Andersson, Charles John 10, 20
 Gate 13, 31
antelope, roan *159*
 sable *159*
anthrax 18, 22, 41, 42, 47
Anti-poaching Unit (APU) 40
Aroe 34
Aus waterhole 15
avocet 57

badger, honey 15, *15*, *101*
beetle, blister *130*
Beisebvlaktes 36
Bloubokdraai 22, 31
Boscia albitrunca 13
buffalo 42
bullfrog, African 36, 37, *135*
Burgers, President T.F. 10
bushbaby, lesser 36, *63*
bushmen 10, *89*
 Heikum 13
bushwillow, red 13
bustard, Kori 32

chameleon *51*, *166*
Charitsaub 13, 26, 29, 42
 lions 26, 27, 30
cheetah 17, 18, 19, 35, 42, 44, 47, 59, 60, 96, *132*, *145*
chestnut, African star 13
Chinese lantern *91*
Chudop 32
Colophospermum mopane 13
Combretum apiculatum 13
 C. imberbe 13
Commiphora glandulosa 13
crow, pied *98*

Damaraland 11, 39
Damaras 20
Diceros bicornis bicornis 39, *73*, *149*
Dik-dik, Damara 13, 22, 24, 31, *130*
disease, control of 41
dragonfly 76
drongo, fork-tailed *145*

eagle, bateleur 20, *120*
 martial 18
 tawny *143*
Ekuma delta 20, 34
 River 10, 35
eland 21, *143*
Elandsdraai 31
elephant *11*, 15, *16*, 18, 19, 23, 24, 32, 39, *39*, 40, 42, *51*, *52*, *102*, *104*, 110, *113*, *119*, *135*, *139*, *140*, *146*, *156*

Enchanted Forest 13, *68*
Etosha
 Ecological Research Institute (EERI) 11, 14, 38, 41
 fencing of 12
 Game Park 11
 Game Reserve Number 1 11, 12
 Game Reserve Number 2 11, 12, 20
 Game Reserve Number 3 11, 12
 geology of 10, 13
 National Geographic article on 12
 Pan 10
 vegetation 10, 13

falcon, red-necked *73*
fencing 12
Fischer's Pan 22, 31, 34, 36
flamingo, greater 22, *136*
 lesser 22, *136*
Fort Namutoni 11, 13
fox, bat-eared 17, 31, 32, 33, *33*, *96*, *149*
 Cape *80*, *143*

galago, lesser *63*
Galton, Francis 10
gecko, Bibron's *109*
geese, Egyptian 28
gemsbok 23, *73*, *153*
Gemsbokvlakte 15, 26
genet, small-spotted 36, *66*
German South West Africa 11
giraffe 14, 16, 23, *32*, *56*, *87*, *88*, *109*, *153*
Goas 26, 28, 31, 32, 42, 47, *82*
Gobaub *139*
goshawk, Gabar 28, 31
Great Onamatoni 11
Greyling, Jan 10
Grootfontein 11
Grootvlakte *150*
guineafowl, helmeted *114*

Halali 10, 11, 13, 17, *17*, 26, 28, 30, 31, 32, 35, 37, 42, 47, 66, *145*, *172*
hare, scrub *60*
hartebeest, red *79*
Heikum 20
Helio Hills 17, 35
Hereros 10, 20
Hirpicium gazanioides 36, *126*, *130*
Hoarusib River 11
hornbill, Monteiro's *82*
 redbilled *82*
 yellowbilled *124*
hyaena, spotted 18, 44, *123*
Hyphaene ventricosa 13

impala 23
 black-faced 18

jackal, black-backed 19, *163*

Kalahari Desert 10
Kalkheuwel 22, 47
Kamanjab 11
Kaoko-Otavi 11
Kaokoland *39*, 40
Kaokoveld 10, *149*
Kaross 21
 Sanctuary for Endangered Species 20, *159*
Karosshoek 21
Koinseb 31

korhaan, black 34
 redcrested 34, *124*
Kruger National Park 38, 41
kudu *12*, 14, 21, 23, 25, 41, *153*, *160*
Kunene River 10, 11, 20

leadwood 13
Leeupan 10
leguaan 29
leopard 23, *23*, 27, 29, 31, 32, 33, 35, 36, 42, 44, 47, *95*, *126*, *129*, *136*, *166*
lion 14, 15, *16*, 19, 20, 25, 26, 27, 28, 29, *29*, 30, 33, *33*, *41*, 42, 44, *65*, *74*, *79*, *82*, *84*, *93*, *99*, *116*, *123*, *154*, *159*, *160*
lizard, agama 28
Lycaon pictus *114*

management, game 11, 12, 38
McKiernan, Gerald 11
migration routes 39, 41, *57*, *150*
mongoose, banded 36
 slender *140*
 yellow *113*
monitor, rock 29
mopane 13
moringa 13
Moringa ovalifolia 13, *68*
Mushara waterhole 52

Namatoni 11
Namatonia 11
Namib-Naukluft Park 38
Namib Desert 11
Namutoni 10, 11, 22, 31, 32, 34, 35, 37, 47, *71*, *143*, *172*
 Fort 22
Natukanaoka Pan 20
Ngami, Lake 10
Ngobib 13, 23, *102*

Odendaal Commission 12
Okaukuejo 11, 13, 14, 15, 16, 22, 24, *24*, 26, 30, 31, 37, 42, 47, *53*, *84*, *145*, *150*, *156*, *172*
Okerfontein 12
Okevi 13, 47
Okondeka 12, 14, 42, *116*, *153*
'Old Girl' 9, 16, 25, *25*, *84*
Olifantsbad *16*, *53*
Ombika 42
Omurambo Ovambo River 10, 34
Omutjamatunda 10, 11
Ondundozonanandana 13
Opononó, Lake 31
Oshigambo delta 20, *34*
 River 10, 35
ostrich 19, *93*, *107*, *116*
Otjitindua 11
Otjovasandu 20, 34
Ovamboland 31, 34
Ovambos 10, 20
owl, scops *119*
Ozonjuitji m'Bari 15, 16, 20, *159*

palm, fan 13
 makalani 13
paper-bark, common 13
plover, blacksmith *109*
 crowned *68*
poaching 39, 40, *40*
porcupine 36, *68*
purple-pod terminalia 13

quelea, redbilled *107*

rabies 41
research 38, 39
rhino, black 15, 21, 28, 35, 39, *39*, 40, 42, *89*, *169*
 desert black 39, *149*
 three-horned black 28, 31, *73*
Rietfontein 10, *11*, 26, 27, 28, 37, 42, 47
 lions 28, 29, 30, 32, *79*
rinderpest 11

Salvadora 12, 13, 26, *94*
sandgrouse, Namaqua *163*
scented thorn 13
Schinz, Hans 11
secretary bird *154*
Selous, Frederick 17
Serengeti National Park 12, 38
shrike, crimson-breasted *130*
sickle bush *91*
spider webs, communal *76*
Spirostachys africana 13
spoonbill, African *63*
Sporobolus salsus 13
 S. spicatus 27
springbok 14, 30, 33, 35, *35*, 36, *36*, 44, *59*, *80*, *84*, *94*, *96*, *110*, *123*, *129*, *132*, *163*
Springbokfontein 12, 47
springhare 36
Sprokieswoud 13, *13*, 37, *68*
squirrel 76
steenbok *124*
Stegodyphus sp. *76*
Sterculia africana 13
 S. quinqueloba 13, 20
stork, Abdim's 33
 marabou *71*, *126*
Suaeda 13, 17, *145*
Suaeda articulata 13
sweetveld 13

tamboti 13
teal, red-billed 28
telemetry, radio 40
Terminalia prunioides 13
terrapin *166*
thornbush savanna 13
tit, Cape penduline 36, 37
tourism 11, 12, 38
Trekkers, Dorsland 10
Twee Koppies 17
Two Palms 13, *120*

vegetation, monitoring of 39, 40, 41
Von Lindequist, Governor F. 9, 11
vulture 19
 lappetfaced *87*

warthog 15, *53*
water thorn 13
weaver, masked *102*
 sociable 15
wild cat, African 34, 36, *65*
wild dog *16*, 17, *114*, *164*
wildebeest *17*, 39, *57*, *75*, *94*, *104*, *126*, *150*

Zebrapomp waterhole 21
zebras *16*, 18, 19, 39, *63*, *101*, *164*, *169*
 Burchell's *80*, *94*, *98*, *99*
 Hartmann's mountain *80*, *98*